FORGOTTEN ORDER OF THE VINEGAROONS

WHIPSCORPION BIOLOGY, HUSBANDRY, AND NATURAL HISTORY

Orin McMonigle

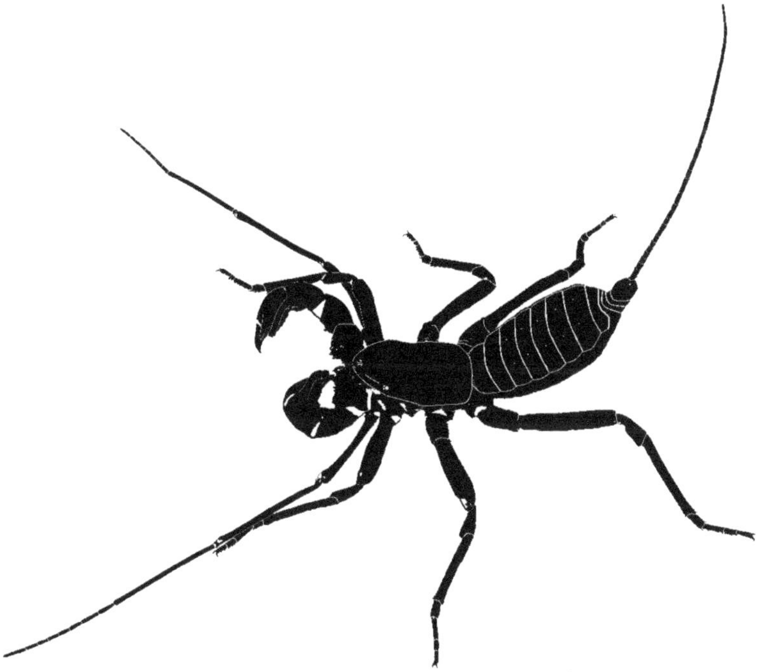

COACHWHIP PUBLICATIONS
GREENVILLE, OHIO

DEDICATION

To my wife Sylvia, also known as Vivian,
whose scary defenses are harmless but flavorful.

ACKNOWLEDGMENTS

I am grateful for the editorial assistance provided by Christian Schwarz, Martin Hauser, Tammy Wolfe, Annie Lastar, Justin Schmidt, Wade Harrell, Jeremy Huff, Sean Sutor, and Christian Elowsky. Thanks are extended to Michael Fanto for multiple year experiences with species cohabitation, specimen, and reproductive data, Shane Burchfield for reproductive data, Trace Hardin for Malaysian *Typopeltis* data and preserved specimens, Cain Eyre for developmental stats, Traci Roach for red form *Mastigoproctus* details, Sam Floyd for collection information, John Lorbiecke for habitat observations in New Mexico, and Martin Gamache for habitat observations in French Guiana. Thanks are extended to my daughters Kree and Gwynevere, especially the latter for showing off her pet. I thank God for providing the resources to make this guide and for creating these amazing gentle monsters.

Forgotten Order of the Vinegaroons

ISBN 1-61646-220-5
ISBN-13 978-1-61646-220-8

CoachwhipBooks.com

CONTENTS

INTRODUCTION

In 2004 I worked on a pair of chapters for a large-format, hardcover textbook that would cover all the arachnid orders kept by hobbyists. It was exciting to be part of this grand undertaking since it was to be a massive text unlike anything seen before. Part of the excitement was familiarity with the main author who had created two well-known color guides to tarantulas that greatly impacted the tarantula hobby in the mid-1990s. My chapters were to cover my two favorite arachnid orders, the Amblypygi and Thelyphonida (at the time Uropygi). I spent many months putting together all the information I had discovered and uncovered over more than a decade, submitted the files, and waited. About a year and a few status requests later, I was informed the publisher had changed their mind. I was very disappointed and tried to forget about it.

At the time I am not certain I had enough information to put together a decent-length stand-alone text combining the two orders, but by the time I considered going in that direction, I had long since lost track of my original efforts. Little desire to start over was another setback. Finally, after a few years, I began to assemble a combined guide that would be called *Whipscorpions and Whipspiders: Culturing Gentle Monsters* (2008). This time, due to limitations of the small guide's construction, I had to leave out some details to make it fit.

From the cover of *Gentle Monsters*.

(Also the pictures had to be small, limited in number, and only a select few in color.) Though compact, *Culturing Gentle Monsters* offered a thorough treatment of whipscorpion husbandry beyond anything seen before. Through Coachwhip Publications I was recently able to expand the whipspider section into a massive, coffee-table sized, full-color text of incredible detail including reproductive data for half a dozen newly cultured species. Unfortunately I had nearly forgotten the vinegaroons. I had not seen a new whipscorpion species imported in six years and the reproductive details had not grown to any great extent in the years between. There have been no verified captive F2 anywhere. However, there is much reproductive data available and I have learned a number of details since that time. Furthermore, I recently acquired a decent size group of *Typopeltis stimpsoni*, a beautiful species with vastly different behavior. I realized there is enough information enabling me to put together a thorough and fascinating text strictly on this order, something that has never been done before, not even in a small format.

This finger placed between the palps of a large female vinegaroon is not being bitten.

CAPTIVE ASPECTS

Whipscorpions combine frightfully amazing looks with a gentle demeanor. Unlike true scorpions, they do not sting and have no venom. They normally will not attempt to bite. Some sources claim whipscorpions are capable of inflicting superficial damage to a person from a nip by the large pedipalps and chelicerae if harassed (Milne and Milne 1980, Stoops and Martin 1995). I have handled numerous creatures from a few different species and have yet to experience even a minor pinch. Sticking my finger directly between the palps or grabbing onto a palp causes nothing more than attempts to flee or rapid flight response. A pinch is not impossible, just highly unlikely during normal handling. Maybe a person would have to try to tear the animal apart or squeeze it in a closed hand to illicit a pinch? The massive pedipalps certainly do look like they could pinch and are opened threateningly, but even the highly defensive *T. stimpsoni* does nothing when a finger is placed between and within the horrifyingly spike-filled mouth. Despite scary looks they truly are gentle giants of the invertebrate world.

In addition to a gentle demeanor, whipscorpions make great pets due to their minimal requirements for housing and feeding. They are long lived compared to most arachnids and require very little effort and minimal expense to keep alive and healthy. Even the arachnid hobbyist just starting out will find them easy to maintain. Inexpensive deli-cups, plastic shoeboxes, gallon jars and small aquariums can be transformed into functional enclosures and take up little room. The majority are content to eat every few weeks. Species that hail from dry forest regions often go without food in nature for as long as eight months every year. They spend the dry season buried underground waiting for the rains. In captivity they feed well on mealworms, crickets, and other commercially available foods which are low cost and readily accessible. (A short section on feeders is included since prey items can be cultured easily and cheaply.)

Whipscorpions have some defensive capabilities, but not ones that could cause more than superficial discomfort to a human. All species hide during the day, which is likely the most effective defense. Some employ a deimatic display to scare off attackers and some turn tail and run away. All species are capable of spraying tiny droplets of defensive acidic secretion. The primary active ingredient is an acid used in food flavorings and salad dressing. This defense usually is not felt by humans unless it gets into a cut or the eyes. However, the defense has great protective value against potential vertebrate and invertebrate predators (McMonigle 2008).

Shedding light on the whipscorpions. Habitat of *Mastigoproctus giganteus* in New Mexico. © John Lorbiecke

The North American whipscorpion hobby relies almost entirely on a single native species, *Mastigoproctus giganteus*, though a few smaller animals like *Thelyphonus caudatus* from Malaysia have shown up in the past from some reptile importers. Our native species has been kept in captivity with some regularity for decades (Frye 1992), but it has always been an exceptionally rare sight at pet shops. Captive-hatched uropygids are produced occasionally. Captive-hatched young have been rarely available but have grown more common in recent years. New tropical species show up every so many years and bring the promise of new established cultures, but it has been difficult to find more than one or two specimens.

Whipscorpions are legal to import and since they are non-venomous, rules and regulations that apply to scorpions are not applicable. They can offer the "scorpion" experiences to those who live in cities or areas where venomous pets are banned. Easily bred tropical species might someday be a real boon to the hobby, but the few that have arrived so far have resulted in a whimper rather than a bang. Unfortunately, while whipscorpions are among the easiest arachnids to keep and rear, development usually takes four to five years and multigenerational reproduction has not been seen (yet). We already have the biggest, most spectacular, and most handleable species, which explains lack of effort on the part of importers and breeders. There is little interest in a smaller, more expensive, but otherwise very similar creature. Imports are also rare because whipscorpions are not found in common export countries like Tanzania. Sometimes, vendors at shows with *Mastigoproctus giganteus* claim the animals are from "somewhere in Asia, probably," because they do not realize such an impressive monster is native. Exotics are difficult to find even as dead specimens. Dried specimens, though obtainable, are rarely seen in collections or for sale and are generally expensive. Liquid preserved specimens of some Asian species had been available in quantity from one biological supply house in the past for anatomy classes.

WHAT TO CALL OUR AMAZING CHARGES?

In this text vinegaroon and whipscorpion will be used interchangeably. One problem with whipscorpion is that it can confuse four different arachnid orders, of which only two are very similar, and yet it is the quintessential term for

Vinegaroons are great for handling and invertebrate displays. 5th instar *M. giganteus*.

the Thelyphonida. Nobody confuses what order vinegaroon refers to, but only because they are not familiar with schizomids. Preference may lean slightly towards whipscorpion since it does not have to be explained to the unfamiliar person. When displaying live invertebrates to the public, the name vinegaroon generates the question: "What is *that?*" Whipscorpion seems to be an easily understood term, but the unfamiliar probably are just recognizing the main word.

"Oh, a scorpion."

Generating curiosity about this strange arachnid that ultimately is very different from a scorpion may be the better course.

The answer, "An arachnid that looks like a scorpion but is more similar to a spider and sprays vinegar in defense," is easy enough.

Whipscorpions are commonly known as vinegaroons in the United States or vinagrillos in Mexico due to the acetic acid component of the defensive spray (the fluid smells strongly like vinegar, which likewise contains acetic acid). There are a few different spellings of the American term *vinegaroon*. Vinegarone is seen referenced in some older texts (Levi and Levi 1968, Frye 1992) but is rarely used or seen. Vinegarroon with two r's is a misspelling sometimes used in the United Kingdom (it is an error rather than an alternate term, since the word originated in the U.S. and there are no vinegaroons in Europe, whether with one or two r's). Sometimes vinegaroon is used only for New World species.

WHIPSCORPION, NOT WHIP SCORPION

The correct spelling is whipscorpion, not whip scorpion. A leaf bug is a bug but a ladybug is a beetle. In the same way a butterfly and dragonfly are not flies while a house fly and horse fly are, the list goes on. A whipscorpion with a space in middle would be a type of scorpion, not a thelyphonid. Some texts use a hyphen in the middle to differentiate the name from the true scorpions, but a hyphen conveys no added meaning and for this reason hyphens are purposefully excluded from most common name schemes except where the meaning would be lost (for example, w-marked cutworm). Rules for invertebrate common name use of contractions and hyphens are not new, but were developed over a hundred years ago by the Entomological Society of America and have been adopted by the American Arachnological Society, the American Tarantula Society, and others (Breene 1995). The problem with a space in the middle is it tells people these animals are members of the Order Scorpiones, not Thelyphonida.

REMEMBERING THE VINEGAROON

Large size and amazing appearance are obscured by the fact vinegaroons are among the least familiar of all arachnid orders. They are among the best known of the least known arachnids, but that just means a single photo or drawing and a couple sentences in a few

Adult male *M. giganteus.*

more general texts. Very few people know a single fact related to the order of the vinegaroons, or even that it exists. The most popular borrowed common name for the order suggests little appreciation of its standing as a unique arachnid order, as different from a scorpion as a spider. Whipscorpions are almost

Typopeltis stimpsoni threatening.

entirely unknown to the general public because they are restricted in range, and even in the right areas are hardly ever seen on accident. There does not appear to be a single mainstream appearance on film, but there is some television footage. In the *Nature* series "Secret Weapons" (1983) episode, *M. giganteus* was shown for almost two minutes. More recently an unidentified vinegaroon showed up for a few unnarrated seconds in a montage on Attenborough's *Micro Monsters* (2013), and a big, handsome male *Thelyphonus* sp. in Laos grabbed just over a minute of footage on *Wild Things* with Dominic Monaghan, in the "Giant Huntsman Spider" (2013) episode. The most extensive footage is probably in the "Tarantula Convention" (2010) episode of *Bugging Out*, where it is featured for slightly over two minutes, sort of. A handsome male Giant Vinegaroon is seen near the end as an aid for therapy to treat arachnid phobia since it is harmless and could be safely touched by the patient despite being monstrous and scary. The only iteration of vinegaroons in popular media is in the online gaming platform *Final Fantasy XI Online* (2002). The "king vinegaroon" is a monster that does not resemble the actual arachnid as it has no whips nor flagellum, but possesses a snake-like neck and head, a forked tail, and a body that sits up high on only four legs. The legs have an entirely unnatural shape and seem to have been copied straight off the main bug troops from *Starship Troopers* (1997), which look nothing like vinegaroons. Only the black color and thick body armor of the "king vinegaroon" resemble the real animal.

The primary reason for putting this text together is to promote appreciation of these fantastic creatures by offering a resource that provides advanced information on whipscorpion biology and husbandry in a readily accessible format that exists nowhere else. There are many dedicated texts on spiders and scorpions, some on mites, two on whipspiders, while even windscorpions, harvestmen, and pseudoscorpions have their own books. There is no such thing as a whipscorpion book. Other than *Gentle Monsters,* what is in print is composed of a few minimally informative paragraphs or pages in general arachnid texts, descriptive papers, difficult to access scientific papers, short husbandry articles from various hardcopy small-run magazines, presentations, non-English articles, or very brief descriptions in

The closest thing to a vinegaroon toy is this glow-in-the-dark keychain.

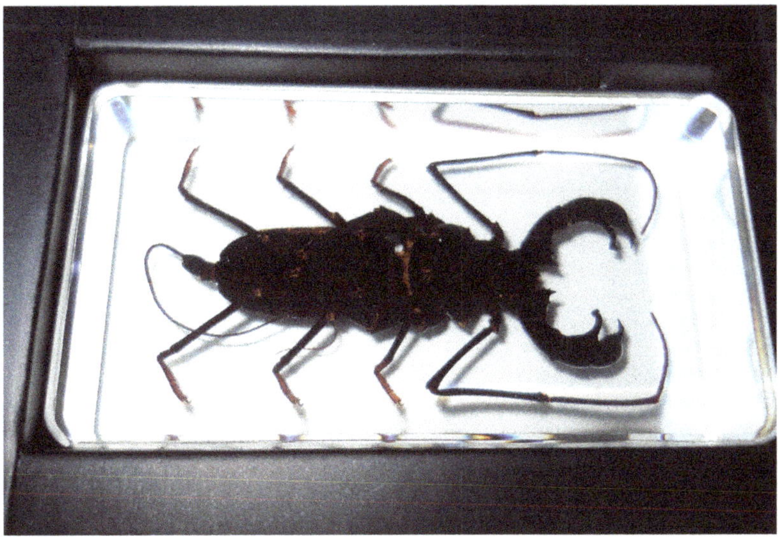

Lucite blocks and keychains with specimens labeled *Typopeltis crucifer* are easily ordered from China. This specimen is a male.

The red mouth of an angry *T. stimpsoni*.

general invertebrate or small animal guides. The spectacular vinegaroons should not be a forgotten order and deserve to have a dedicated text.

The best efforts and hours of research have gone into finding the most recent classification information for the whipscorpions in this handbook. However, taxonomy is difficult since external characteristics can be lacking and make it difficult to identify specimens below the generic level. The lack of concentration on this group, along with papers changing the classification through phylogenetic studies, may make the scientific names somewhat fluid for the foreseeable future. There are still fifteen genera as listed in *Gentle Monsters*, but four were removed and four added. Four Asian genera, *Abaliella, Chajnus, Minbosius*, and *Tetrabalius*, were synonymized with *Thelyphonus* (Haupt 2009) and a new genus erected by Krehenwinkel et al. (2009) for a new Philippine species. One New World genus, *Ravilops*, was created for a species from the Dominican Republic, and two new genera, *Mayacentrum* and *Valeriophonus*, were erected for Central and South American species formerly from the genus *Mastigoproctus* (Viquez and Armas 2005, 2006). The subtitle of this book is whipscorpion biology rather than uropygid or thelyphonid biology because arachnologists cannot agree on the name of the order (Harvey 2003).

It is difficult to talk about whipscorpions without addressing the names of various body parts. Standard arachnid terminology, labeled figures, and a glossary are provided to help the uninitiated who may find terminology initially daunting. Some confusing aspects of terminology are disagreements among arachnologists on palpal segmentation relating to the position of the apophysis (which is really a spine or finger since an apophysis by definition should be a blunt-tipped attachment point for muscles), and the abdomen which is the opisthosoma having a pygidium or postabdomen which is analogous to the scorpion metasoma, and yet conservation of terms is seldom practiced. Lack of concentration on this order seems to be the cause for lack of consensus though some discrepancies between various works affect the entire class.

The purpose of this book is to present all the available information in a format that is worthy of these strange, oddly beautiful, and amazing arachnids. The goal is to combine a husbandry manual, a picture book, and a biological textbook tailored to the advanced arachnid enthusiast or nature buff. Hopefully the availability of useful husbandry details will encourage expanded captive-breeding efforts, research, or even conservation (if one person could be convinced not to go out of their way to run these over on the road at night or to rescue rather than smash specimens at a construction site). Hopefully the photographs will entice those who are considering keeping and learning about these spectacular pets. This handsome arachnid order contains truly amazing animals that provide superior display specimens and easy captives. *Forgotten Order of the Vinegaroons: Whipscorpion Biology, Husbandry, and Natural History* provides detailed information for the successful husbandry and understanding of these creatures.

Typopeltis eating *Popillia.*

1
OTHER WHIPS AND SCORPIONS

Class Arachnida
- Order Thelyphonida (Uropygi)—Whipscorpions
- Order Schizomida—Short-tailed Whipscorpions
- Order Amblypygi—Whipspiders (Tailless Whipscorpions)
- Order Araneae—Spiders
- Order Palpigrades—Microwhipscorpions
- Order Scorpiones—Scorpions
- Order Pseudoscorpiones—Pseudoscorpions
- Order Solpugida (Solifugae)—Windscorpions
- Order Opiliones—Harvestmen
- Order Ricinulei—Hooded Tickspiders
- Order Acari—Mites and Ticks

There are eleven extant orders of Arachnida, including the familiar spiders and scorpions. Four of the arachnid orders include the term whipscorpion in their common name because the front legs are modified into elongate tactile organs that function like the antennae of other arthropods. Beyond these four orders and scorpions, another two orders include the word scorpion. Some workers in the past have tried to roll the Amblypygi, Thelyphonida, and Schizomida together into a single order, the Pedipalpi, but there is little support for this idea. Even before DNA analysis the defining characteristics were not very convincing (such as missing tarsal claws that are not actually missing and mirrored in some unrelated taxa (Harvey 2003), or "large palps"). Recent molecular studies have found the Amblypygi is actually the sister taxon to spiders (Wheeler and Hayashi 1998, Weygoldt 2000) including similarities in the primary nervous system and sperm (Ax 2000).

Whipscorpion flagellum and whips.

Developmental analogies can be telling, but usually are not a major tool in understanding relationships—a live-bearing fish is no more closely related to a live-bearing cockroach than to an egg-laying one, nor is a centipede that adds segments as it grows more closely related to a millipede than to a centipede which does not. Though there is some correlation to brooding

1st instar whipscorpions must cling to the mother`s opisthosoma. © Michael Fanto

Pseudoscorpiones, Scorpiones, and early development of Araneae, reproduction and early development is most similar for the orders Amblypygi, Schizomida, and Thelyphonida. Maternal care is extremely important and is required until immatures reach 2nd instar; maternal care may continue on much longer, but after that period is no longer life or death. Eggs develop inside the female's opisthosoma for a few months and then are extruded inside a membranous, fluid-filled sac (the eggs are never exposed to air). Inside this broodsac they are nourished, or at the least hydrated, by fluids from the female's oviduct until the postembryos are ready to hatch into 1st instars. Newly hatched 1st instars cling to the female's body until the first molt and tend to encircle the opisthosoma (different from scorpion 1st instars that stay only on top of the mother's opisthosoma). At this stage they have limited mobility, underdeveloped appendages, lack protective coloration, and do not feed. Once they molt into 2nd instars they look like miniature adults, are capable of surviving beyond the mother's opisthosoma, and are ready to hunt for food on their own once the exoskeleton has hardened. (This is normally the stage when spiders leave the eggsac to feed.) Immatures will eat live prey and molt a set number of times (commonly four to six) before reaching sexual maturity. After reaching maturity they live multiple years and females tend to produce a single, often seasonal, brood each year. Adult scorpions, pseudoscorpions, whipscorpions, and short-tailed whipscorpions do not molt after adulthood, but all whipspiders and some female spiders continue to molt.

The uropygids or whipscorpions were recently renamed as the order Thelyphonida because the Uropygi originally included thelyphonids as suborder Holopeltida and schizomids as Schizopeltida in reference to the incompletely fused carapace of the latter (schizo- means "split"). Although the schizomids were elevated to an order, uropygid is still an accepted common name for thelyphonids. It is rarely used because it was never commonly used, and use of the name in classification is somewhat contentious. Uropygyi is sometimes elevated to a superorder containing these two orders.

Whipscorpions are more closely allied to spiders than scorpions. The body is elongate and superficially looks like a scorpion, but there are major structural differences. The chelicerae are two-segmented like in spiders and do not end in a claw like the three-segmented chelicerae of scorpions. The pedipalps terminate in small, opposable tarsal segments that resemble the scorpion pincer, but there is an additional large spine (apophysis) on the tibia. The abdomen (opisthosoma) is connected by a narrow waist to the prosoma as seen in amblypygids and spiders, but it is wider dorsally, including a tergite. Although technically within the widened preabdomen (mesosoma), vinegaroons have two pairs of book lungs in the same position as primitive spiders and whipspiders, which do not have mesosomas only because they have no metasomas. (Scorpions have four pairs of book lungs, one pair each on mesosomal segments three to six.) The abdomen ends in a tubular, fully articulate pygidium or postabdomen. The pygidium is similar in structure and movement to the scorpion's meta-soma, but consists of three rather than five abdominal segments. The pygidium ends with a whip-like flagellum rather than the true scorpion's telson. Glands in the opisthosoma that open on the pygdium near the base of the flagellum produce and secrete the defensive fluid, while the flagellum helps to direct this fluid towards the threat. In scorpions, the venom glands are not located within the abdomen but inside the telson. The whipscorpion's large median pair of eyes is located at the front of the prosoma against a raised ridge similar to a scorpion's, and is not located on a defined plate (or ocularium) like in amblypygids and most harvestmen. The two lateral clusters of three eyes (five on a single known species) are located further back to the sides. (In scorpions the lateral eyes are forward of the median eyes.) There are 110 species of whipscorpions (Prendini 2011d) worldwide with a majority of described species from Southeast Asia. Unlike scorpions, which vary greatly in color and shape, there is only one family and very little deviation in form. The largest species ever kept in captivity, and arguably the largest in the world, is a common U.S. native, our only native species, *M. giganteus*. Large specimens are up to 3″ (76 mm) in body length, and nearly double that when the flagellum is included. Inclusion of the flagellum seems to be what leads to tall tales (not tails) of "unnamed exotics" reaching 6″ (152 mm). Fortunately the myth does not double again in size every so many years.

Scorpions are outfitted with palps that terminate in gigantic tarsal pincers, three-part

Wild *Smeringurus mesaensis* scorpion from Arizona. Compare body structures to the whipscorpion.

chelicerae each ending in a claw, four walking leg pairs, and a remarkably specialized opisthosoma ending in the telson, which is used to sting prey and predators. The opisthosoma is divided into the wide seven-segmented mesosoma, which contains book lungs and reproductive organs, and the articulate, tail-like metasoma that ends in the telson. The telson includes the stinging apparatus of vesicle

Scorpions come in various colors, unlike whipscorpions. *Rhopalurus junceus*, captive-born immature.

(venom bulb) and aculeus (stinger) and is unique to scorpions. It is not part of the abdomen, which ends with the fifth metasomal segment. Tails are often ignored in the body measurement of arthropods, but no matter how thin and appendage-like the scorpion tail (metasoma) is, it is part of the abdomen in a way something like the flagellum of whipscorpions is not. Inclusion of the elongate, skinny metasoma of male *Hadogenes troglodytes* can give the longest specimens a 9″ (23 cm) body length (legspans are shorter). A less familiar and unique trait of scorpions is possession of trademark appendages, called pectines, located behind the rear legs. The comb-like pectines are organs used to detect vibrations, surface texture, moisture, odors, and possibly pheromones. The median eye pair is not located on a distinct plate and the lateral eye clusters vary in number and arrangement (zero to five each) in a line rather than a cluster. The exoskeletons of all known scorpion species fluoresce under ultraviolet light, while whipscorpions do not. There are around 1,500 species of scorpions in 13 families that range widely in size, shape, and color. Of these only about two dozen have venom that is lethal to humans. However, some humans can become allergic to and die from any scorpion venom.

Uroctonus mordax 1st instars hang onto the mother`s opisthosoma, similar to the whipscorpion.

Scorpion reproduction is similar in some ways to whipscorpions, but females are live bearing. Males produce spermatophores taken up by the female during a courtship dance where the pair holds “hands” (pedipalp pincers). Females retain developing eggs within the abdomen and give birth to postembryos that immediately hatch into 1st instar immatures. The 10-100 eggs are retained in the mesosoma and nourished internally until birth. As with whipscorpions, a lot of time and energy is invested by the female in a single brood. Eggs are never exposed to air and time from mating to 1st instar is rarely less than six months (McMonigle 2009). First instars are pale, do not feed, and cling to the mother’s back even after molting to the next instar. Second instars are pigmented and capable of feeding but commonly remain on the female's opisthosoma a few weeks or more following their molts.

Hubbardia from California, under a rock in a backyard garden. "There was only one when I spotted this one, but when I returned to see if there were more, I found another three. There`s an apparent association with these and fire ants, but I haven`t seen any fire ants on our property." No additional specimens have been spotted between Dec. 30th, 2012, and Sept. 2013. © Cynthia Bingham Keiser

The closest relatives of whipscorpions are not scorpions but representatives of the Order Schizomida (short-tailed whipscorpions). The two orders are sometimes placed together in the Superorder Uropygi. There are two families and 260 species worldwide (Prendini 2011a). Schizomids are neat little creatures that rarely exceed 1/3" (8 mm) in length and are similar in mass to pseudoscorpions. Since they are so tiny some people confuse them with microwhipscorpions but palpigrades are even smaller. Schizomids look very much like miniaturized whipscorpions except the flagellum is short and the male's pygidium ends in a small flattened or globular structure related to reproduction. (This looks more like the scorpion's telson than the whipscorpion's flagellum.) Less notable differences include vertical rather than horizontal palp orientation, lack of lateral eyes (most have no eyes at all), a single pair of book lungs, and a carapace whose segments are incompletely fused towards the back. There are no available test results on the defensive secretions but the smell is similar to vinegar. Females hold the eggs in an external sac like whipscorpions, but with far fewer eggs. Less than ten young hatch after about three weeks (Sejna 2005). They also develop through five instars. Adults do not molt and can survive a few years in captivity. Nearly all species are found in tropical areas and are uncommon, though they supposedly can show up in greenhouses from time to time. There are a few U.S. species but none with such fanciful scientific

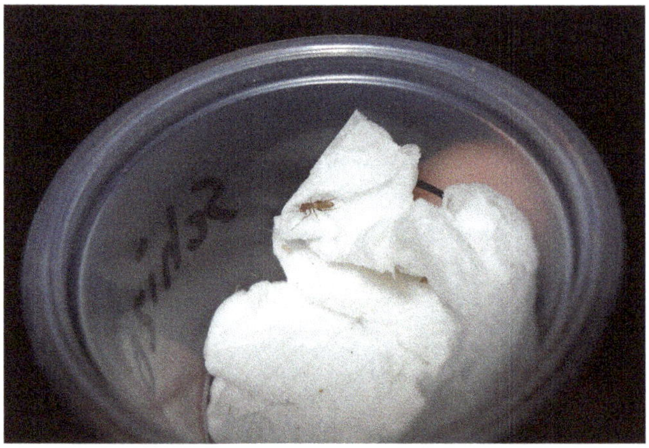

Stenochrus portoricensis size compared to a 1oz. condiment cup. Schizomids do not get a lot bigger.

Close-up of *Stenochrus portoricensis*.

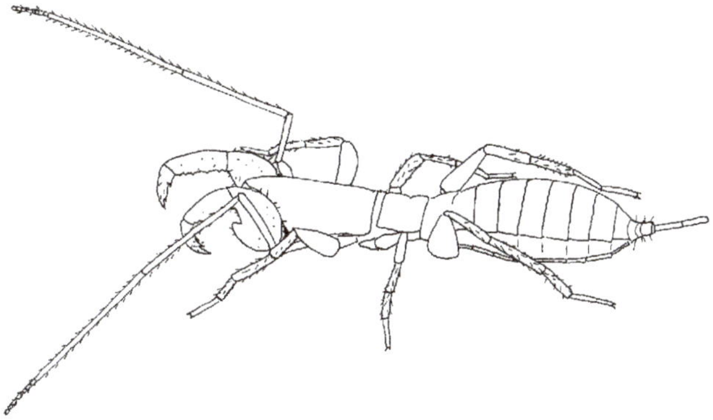

Shizomid females have a short, thin flagellum.

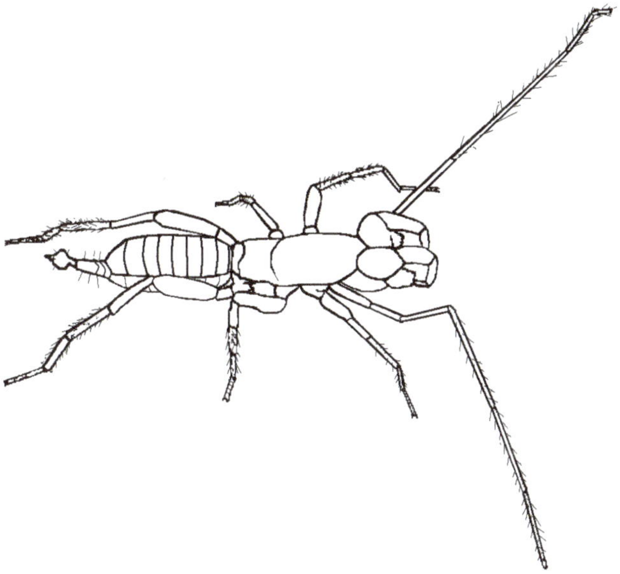

Shizomid males possess a globular or flattened flagellum that can look a little like the scorpion telson.

names as the Australian species known as *Draculoides bramstokeri*. According to Levi and Levi (1968) at least one species is found in southern California, *Hubbardia pentapeltis*. There is also a tiny parthenogenetic species found in Florida, *Stenochrus portoricensis*. Other schizomids are capable of reproducing through parthenogenesis, but *S. portoricensis* is among the few that only reproduces in this way (no male has ever been seen). Although a few specimens of Floridian *S. portoricensis* were seen on a single vendor list in mid-2013, a person interested in studying these might still have better chances of successful acquisition through flying to the tropics to collect and obtain stock. The short-tailed whipscorpions are rarely kept because specimens are nearly impossible to locate, difficult to observe without magnification, and can be very difficult to keep alive.

The Amblypygi (whipspiders) share some characteristics with vinegaroons, but close inspection reveals many are not even cuticle deep. Both orders have grabbing mouthparts that are so large they appear to be an oversized pair of front arms. The prominent mouthparts (pedipalps) are usually folded up, are normally much more massive than the walking legs, and are important for mating. However, the palps are not fused at the coxae in amblypygids and they do not end in a pincer. At least one popular general entomology encyclopedia (O'Toole 2002) claims the whipspiders also have a pincer formed by the last two segments of each palp, but of course the author must not have looked at an actual specimen. Both orders have

Whipspider palps do not end in a pincer. *Damon diadema* hand.

TABLE 1: DIFFERENCES BETWEEN THELYPHONIDA AND AMBLYPYGI

	Whipscorpion	Whipspider
Chemical Defense	Yes	No
Flagellum	Yes	No
Pygidium (Metasoma)	Yes	No
Sucking Stomach	No	Yes
Prosoma	Long, narrow	Wide, short
Median Eyes	Located on a central ridge like in scorpions	Located on a raised plate or ocularium
Whip Patella	Extremely long	Extremely short
Whip Articulation	Tibia unarticulated, tarsus 8-9 articles	Tibia and tarsus greater than twenty articles each
Autotomy	Little to none	Advanced
Regeneration	Limited	Advanced
Molting	Substrate surface	Hanging
Growth	Five instars	Six to eight instars
Postultimate Molt	Never	Two or more
Sociality	Limited	Prolonged subsocial
Burrowing Ability	Excellent	None
Gestation	Inside sealed burrow	Free moving
Infertile Egg Production	Yes	No, with exception of rare parthenogenetic species
Broodsac	Partly attached	Fully attached and edged by pleural folds
Palp Tarsi	Pincer	Stiletto
Palp Orientation	Horizontal movement	Horizontal and vertical movement
Basic Movement	Forward	Multidirectional
Leg and Palp Form	Uniform	Variable
Palpal Coxae	Fused	Separate

the characteristic eight legs of arachnids, but there are only six walking legs since the front legs are useless for locomotion and function like antennae. Although both orders have highly specialized antenniform forelegs, so do some mites, windscorpions, and palpigrades. The amblypygid antenniform leg structure is highly attenuate and extremely different in segmentation. Instead of being massive and elongate, the patella is hard to tell from a tiny tibial article while the incredibly thin and long tibia and tarsus are comprised of a hundred or so tiny articles. The antenniform legspan of some species is known to reach twenty-four inches (61 cm). Whereas all early instar amblypygids have disruptive coloration in the form of white markings on the whips, this is not known for any species of whipscorpion. The last three leg pairs are built for rapid sideways movement rather than for forward or backward locomotion. Whipspiders have a compact, round body shape which superficially resembles a spider's, likewise with a very narrow waist without a dorsal tergite. The armored, segmented opisthosoma is very different from the membranous opisthosoma of all but the primitive Liphistiomorph spiders. A few have reduced or missing eyes but even cave species usually have eight eyes; two are located at the middle front of the prosoma and a cluster of three eyes is found on each side. The eye clusters on the sides face upwards, are closer to the front, and more noticeable than on whipscorpions. Amblypygids have a much reduced tail segment at the end, but it never ends in a flagellum like the whipscorpions and there is no metasoma (no pygidium). This order has proven somewhat easy to rear in captivity through multiple consecutive generations and some of the largest species can pass through six molts and reach maturity in less than ten months (McMonigle 2008, 2013). Captive-hatched and captive-born are common for the few available species. There are presently 161 described species in five families and seventeen genera (Prendini 2011c).

Whipspider median eyes are located on a notable ocularium, *Heterophrynus batesii*.

On the microscopic level, the whipspiders and whipscorpions are very different on almost every surface and structure, especially when it comes to the sensory apparatus of the antenniform legs. Whipspiders are outfitted with porous, rod, club and leaf-like setae of all different sizes and lengths used to detect texture, movement, shape, humidity, vibration, odor, and air movement, whereas whipscorpions have a relatively limited sensory apparatus and seemingly just a single type of tactile setae. To illustrate the overall difference on a microscopic level, the technician who helped with the whipspider Scanning Electron Micrographs actually thanked me for the opportunity to see such magnificently structured and variable specimens. After the whipscorpion SEMs there were no thanks, just a comment about them being simply dirty and the structures too large. I do not agree with the assessment (a tank can be just as pretty as a sports car in the right eyes) but I understand the point of reference (and the images did require a bit of cleaning).

Despite some resemblance, there is a long list of differences between the whipspiders and whipscorpions. Some of the more notable and easily observed differences are contrasted in Table 1. After reaching maturity both live multiple years and females can reproduce seasonal broods. However, there is some fluidity in whipspider maturity with no parallel in whipscorpions. The greatest behavioral difference related to body structure is that amblypygids must rely on existing shelters while uropygids can construct their own. (Surprisingly in opposition to this great advantage, whipscorpions are comparatively limited in range and numbers.) The whipscorpions, like the true scorpions, have thick palps adapted to be formidable digging tools.

Whipsiders must hang to molt properly.

Microwhipscorpions, or Palpigrades, look a lot like pale, white, nearly microscopic versions of their larger cousins but are no more

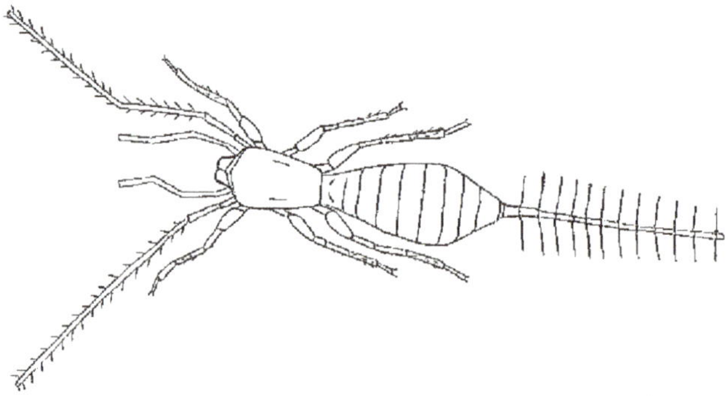

Palpigradi.

closely related to them than are spiders. They use the palps as an extra set of legs and have three-part chelicerae like Scorpiones and Opiliones. Palpigrades can be truly microscopic and make the Schizomida seem huge; adults range from 0.67 mm to 2.8 mm in body length (Beccaloni 2009). They are subterranean, mostly tropical, and generally rare. Specimens have not been maintained successfully in captivity and the means of reproduction remains undocumented. Their pedipalps are very different and look more like walking legs than like the crushing, powerful weapons of the other "whipscorpion" orders. They are mostly tropical like all the small orders, but a few species are found in the southwestern U.S. Not only difficult to find and restricted in range, near-microscopic size is an added detraction. There are around eighty known species (Harvey 2003).

Windscorpions (Solpugida or Solifugae) are around the same size as whipscorpions and are also excellent burrowers. Except for a small group of highly adapted burrowers, they have long legs and are built for speed. When chasing down wild *Eremobates* in Arizona, the reason for the common name becomes obvious; they move like leaves blown rapidly across the ground on a particularly windy day. Windscorpions can be quickly differentiated from whipscorpions because they have unique

Windscorpion and burrow, *Galeodes* species.

Pseudoscorpions are tiny. Females carry a broodsac like whipscorpions. (The eggs are not separate but are encased in a single, tight-fitting membrane.)

pedipalps and massive chelicerae that are usually larger than the head. Also, the cephalothorax has a distinct head and thoracic region, so together with the abdomen there are three visible body regions. The head and thoracic segmentation is structurally and visibly very different from the incompletely fused schizomid prosoma. Windscorpions are outfitted with long, notable setae and baton-shaped palps that end in a sticky, retractable organ which aids in prey capture and allows some species to climb smooth glass. (Such climbing looks like swinging from monkey bars and can be shockingly rapid in function.) On close inspection the

Captive-hatched windscorpion third instar feeding on a tiny lobster cockroach. *Rhagodes* species ex. Egypt.

front pair of legs appears visibly thinner and has reduced or missing tarsal claws because they are not used for walking but serve a primarily sensory function. The sensory front leg pair is incredibly reduced compared to the ambulatory legs in some burrowing species. Despite the obvious specialization in form and function, the front legs are not referred to as antenniform legs because they often contact the ground. Windscorpions do not make webs and the legspan of the largest species does not exceed six inches (15 cm). On the bases of the rear pair of legs are unique, mallet-shaped organs called "racquet organs" which perform a similar function to the pectines of scorpions. Males have accessory organs on the chelicerae or practice direct fertilization. Females dig a burrow and lay 40-200 eggs in a pile, but these rarely hatch in captivity and rarely continue past a few instars if they do. There is no visible postembryo stage (for *Eremobates* spp. at least), as eggs hatch into 1st instars that are functionally similar to the 1st instars of Amblypygi or Scorpiones. They have developed setae and appendages, and are capable of movement, but are sedentary and do not feed

Some mites have a pair of elongate legs used as antennae such as the specimen on the right. These two different, large mite species spend their entire lives on a big millipede, *Acladocricus* sp., from Negros Island, Philippines.

until the molt to 2nd instar (McMonigle 2009). There are 1,113 species (Prendini 2011b) found primarily in desert habitats. Only a single species from India is known to possess venom glands, but they are not functional (Punzo 1998).

Pseudoscorpions, as the name implies, look a little like scorpions from the palps to the middle of the body, but there is no metasoma, no telson, nor a flagellum. The chelicerae are two-segmented like in whipscorpions, and the venom is located in the claws of the pedipalps. These tiny creatures breathe through trachea and are more like windscorpions than scorpions. Like spiders they can spin silk, but with the chelicerae rather than abdominal appendages. Representatives of the same species live in close proximity and are tolerant of conspecifics, but will rapidly hunt down and attack other species in the same container (McMonigle 2009). Males convince females to take up the spermatophore through a courtship dance, complex webs, or unknown means. Females produce up to a few dozen eggs that hang from the abdomen in a brood pouch similar to whipscorpions, but postembryos are nourished extensively inside the broodsac and 1st instars

Many mites have enlarged sensory front legs that are outfitted with long setae like this specimen, but they can be difficult to see even when magnified. © Henry Kohler

do not cling to the mother's back. Females only produce young once or twice a year. They are widespread and are sometimes found in greenhouses or spread by man like the short-tailed whipscorpions. There are twenty-six families with more than 3,454 species (Harvey 2011). Although most, if not all, are venomous, they are too small to puncture the surface of the skin and cannot deliver a noticeable bite to humans.

The orders Opiliones and Ricinulei have little to do with whipscorpions except as possible prey items. Mites from the order Acari are sometimes a problem as cage pests.

Some arachnids undergo truncate and rapid early development such as these Opiliones. 1st instars hatch directly from the egg and molt within minutes to 2nd instars.

Whipscorpion, Chengelpattu, India. © Sandrine D. Martinez

Typopeltis stimpsoni uninterested in finger food.

2
BIOLOGY AND MORPHOLOGY

THE MAJORITY of whipscorpion species are found across Asia and warm areas of the Americas. They range in size from just under three quarters of an inch (17 mm) to three inches (76 mm). There is only a single extant family, the Thelyphonidae. No representatives of this order are found in Europe, Australia, New Zealand, or Madagascar. A single species, *Etienneus africanus*, is native to western Africa, though it was originally thought to be an accidental introduction from Asia (Huff and Prendini 2009). The only species found in North America, north of Mexico, is the gigantic *Mastigoproctus giganteus*, whose largest specimens are the biggest in the world. The New World genus *Mastigoproctus* contains the largest and had included some of the smallest species. *Mastigoproctus maximus* from Brazil may grow as large as the largest *M. giganteus*, while *Mayacentrum* (*Mastigoproctus) tantalus* from Central and South America is among the smallest at 17-21 mm from the front of the prosoma to the end of the opisthosoma (Viquez 2003, Viquez and Armas 2006).

Overall color and shape of the various species is quite similar. Like a Model T Ford, the whipscorpion carapace comes in any color you want as long as that color is black. Intensity of the basic black fading to brown is the only common variation. There are a few known aberrations and some species in which the black fades partly or entirely to a reddish brown, though much of the red color seen in photographs is due to overexposure bringing out color or reflection not seen in normal lighting. The tarsi are commonly red-brown. The joints, intersegmental and pleural membranes can be black, tan, or white. The 2nd instars of some species have bright red pedipalps, but the red changes to black as they develop through molts. The body structure is highly uniform across genera and has changed very little from fossil species (Edgecombe 1998). Species from around the world look incredibly similar to our native *Mastigoproctus giganteus* but are told apart by somewhat subtle differences in the spines on the pedipalps, ridges on the prosoma, tibial spurs, or internal female genitalia. Even the pedipalps, antenniform front legs, and walking legs vary little from species to species. A serious enthusiast may be thrilled by minor differences in shading and spine or ridge shape, but most general arachnid hobbyists would be unlikely to discover differences that are not pointed out. No other arachnid order is so amazingly uniform in construction. On the other hand, truly fascinating variation can lay in the behavior and developmental requirements. Reproductive periods and nutritional requirements might also be widely variable. Lack of significant structural differences is why there is only a single family. This includes just four subfamilies and fifteen genera.

The bright red palps of 2nd & 3rd instar *M. giganteus* may be warning coloration.

The palps of 2nd instar *Thelyphonus* (*Tetrabalius*) sp. ex. Borneo are black like the rest of the body. © Paul Bollinger

This small (2nd or 3rd instar) *Typopeltis stimpsoni* also has dark pedipalps.

Family Thelyphonidae
- Subfamily Hypoctoninae
  - Genus *Etienneus*
  - Genus *Hypoctonus*
  - Genus *Labochirus*
  - Genus *Ravilops*
  - Genus *Thelyphonellus*
- Subfamily Mastigoproctinae
  - Genus *Mastigoproctus*
  - Genus *Mayacentrum*
  - Genus *Mimoscorpius*
  - Genus *Uroproctus*
  - Genus *Valeriophonus*
- Subfamily Thelyphoninae
  - Genus *Ginosigma*
  - Genus *Glyptogluteus*
  - Genus *Thelyphonoides*
  - Genus *Thelyphonus*
- Subfamily Typopeltinae
  - Genus *Typopeltis*

INTRASPECIFIC INTERACTIONS

Where related to mating, intraspecific interaction is covered in the reproduction chapter. Whipscorpions are usually considered solitary animals and captive cohabitation can lead to unnecessary losses even when there is plenty of space and food. Breene (2000) reports four or more 2nd instar *M. giganteus* have been found living together under rocks in Carlsbad, New Mexico, and Schmidt et al. (2000) also noted wild specimens were occasionally found together beneath rocks, but there are plenty of cannibalistic predators sometimes found under the same rock or on the same plant. In captive situations, female *M. giganteus* will share captured prey with their offspring for the first few months and usually do not cannibalize at this time (Schmidt 2009). In captivity, *M. giganteus* sibling groups have been kept together for as long as three years without cannibalism and may share burrows (though never around the time of the molt), while some of the Asian *Thelyphonus* have been kept communally without cannibalism as adults (M. Fanto pers. comm.). Inversely, immature *Thelyphonus* seem to be cannibalistic. Of course a decent enclosure with appropriate shelters and adequate feeding is required even in cases where animals could possibly cohabitate peacefully.

BEHAVIOR

Those who have only seen *M. giganteus* in action may think whipscorpions are all lethargic and methodical in their movements. Animals can be prodded, poked, or grabbed with a hand and they tend to, rather slowly, mosey away from the scene of disturbance. Lack of reactive movement sometimes to the point of lethargy is so unusual that researchers have

2nd instars will share meals (or at least fight over and not kill each other). © Sara Pagel

Older immatures would rapidly kill and eat each other if left together.

These mature females are trying to kill each other and must be separated right away.

This male and female *Thelyphonus caudatus* are attempting courtship, not fighting. © Michael Fanto

even worked out theories on how the construction of the legs must prevent them from walking at a faster pace (Shultz 1992). Although *M. giganteus* are capable of shocking bursts of speed while hunting, such episodes seldom involve running and are uncommonly seen. The small Malaysian *Thelyphonus caudatus* are a faster moving animal but still could not be considered speedy by arachnid standards. The large Japanese *Typopeltis stimpsoni* is a skittish creature that runs rapidly or springs into a defensive stance when disturbed. Large specimens in flight mode are capable of building up enough speed within a second to push themselves up over the sidewalls of a three-inch tall deli-cup. Unfamiliar specimens of this species transferred into a small terrarium together can chase each other around at a decent pace. As soon as the antenniform tarsi touch, one or both run until they hit the wall. It takes barely more than a second for one to move six inches.

A limited deimatic display is seen every so often in captive *M. giganteus,* but is usually in response to another small animal rather than a human. When seen it often consists of simply opening the palps wide. On the other hand, the threat display of *Typopeltis stimpsoni* is expressive and consistent. The palps are opened wide in warning and expose a bright reddish mouth (striking colored setae on the chelicerae). At the same time the opisthosoma is reared up and the pygidium swings forward to direct the flagellum towards the threat. As common to deimatic displays, only large immatures and adults show this behavior, since tiny animals have relatively little to display and have hardly anything to back up the threat. *Thelyphonus caudatus* does not seem to have any sort of deimatic display even at maturity.

Animals recently taken in the wild are more aggressive than those kept in captivity even for a short period. I have never had an animal

Deimatic display in *Typopeltis stimpsoni.*

pinch me, nor seen an honest attempt, and other keepers have reported the same. However, a few collectors report a slightly modified story as represented in the following quotes.

> "I have had some give me a good pinch, hurts a little when the point of the palps gets under your fingernail. It only seems to ever happen right in the field and then afterward they are not so vicious" (Sam Floyd pers. comm.).

> "I have caught a bunch of them in the wild. They act very different when confronted in the wild. It is odd how they calm down so fast in captivity. They just do not act like they do when you first come across them in the wild. They lunge the best they can and try to pinch, but not much with each individual pedipalp, they try to 'pinch' like you see them

> catch prey, with both, bringing them together. What they do in the wild is lunge, do the pinch motion real fast and spray all at the same time, then they turn around and make a run for it! After that I can pick them up and handle them all the time, never seeing that behavior again. But I have never been pinched, I have only seen them try" (D. T. Anderson pers. comm.).

Rapid loss of aggression in captivity is not common but also not unheard of. Freshly caught millipedes such as the large Floridian *Narceus gordanus* writhe terribly like a fish out of water but following a few weeks in captivity this behavior is never seen again.

Most activity is restricted to after the lights go out, but if all light is cut off in the middle of the day they do not necessarily become active. Whipscorpions can be observed with a flashlight at any time since they are active till late morning, slowly probing with the antenniform legs or moving around dirt. A flashlight will disturb the very rare grooming behavior and sometimes causes an animal to retreat into its burrow. Daytime behavior can be forced by adding active prey or other whipscorpions to a cage. Feeding, mating, and fighting are difficult to disturb unless the animals are picked up and forcibly separated.

Burrowing behavior can vary according to season, habitat, and species. Species found in seasonally dry areas, such as *M. giganteus*, dig a deep burrow in the ground and are impossible to find most of the year except for a few months following seasonal rains. However, the same species in the same general area when a population is found near water sources (even surrounding man-made pools and cement spillways) are active and can be found for a greatly extended season (Lorbiecke pers. comm.). During their active hunting season they do not waste energy creating extensive burrows and prefer to use whatever is available or requires little modification. They still hide during the day but can be found in somewhat shallow and existing burrows or excavate shallow scrapes under rocks. In captivity the larger specimens tend to dig deep burrows in various shapes while small species and small immatures tend to scrape out some dirt under small pieces of wood or plastic. Even large specimens are thigmotactic (a quick way to say they show a marked preference for shelters that touch the top and bottom or all sides of the body) but the construction of burrows is elaborate.

This subadult male *Typopeltis tarnanii* is cleaning its whip with the palps. © Cain Eyre

Whipscorpions seem to be very inquisitive creatures. The desire to probe and test is especially evident when placed in containers with punched holes they can stick the antenniform legs through. They react differently to prey, fruit, or human fingers. A desire to test and touch objects would seem to be of little value if they could not process or remember. Punzo (2005) ran various tests on *M. giganteus* and

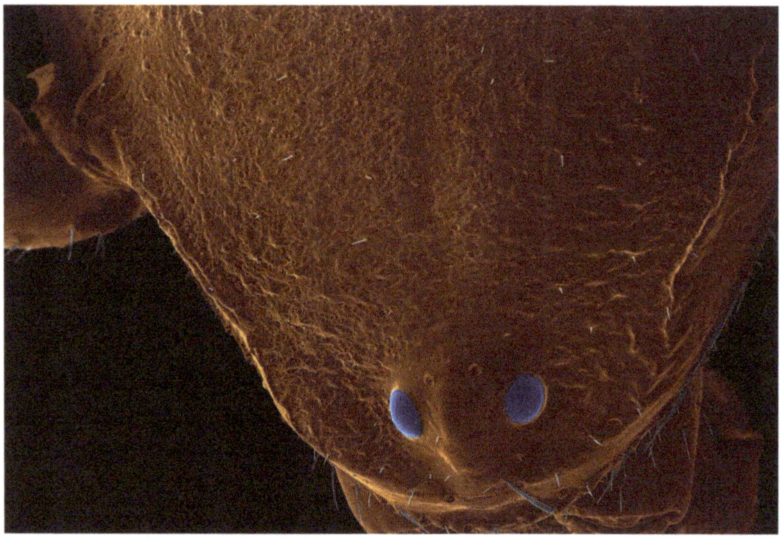

T. stimpsoni.

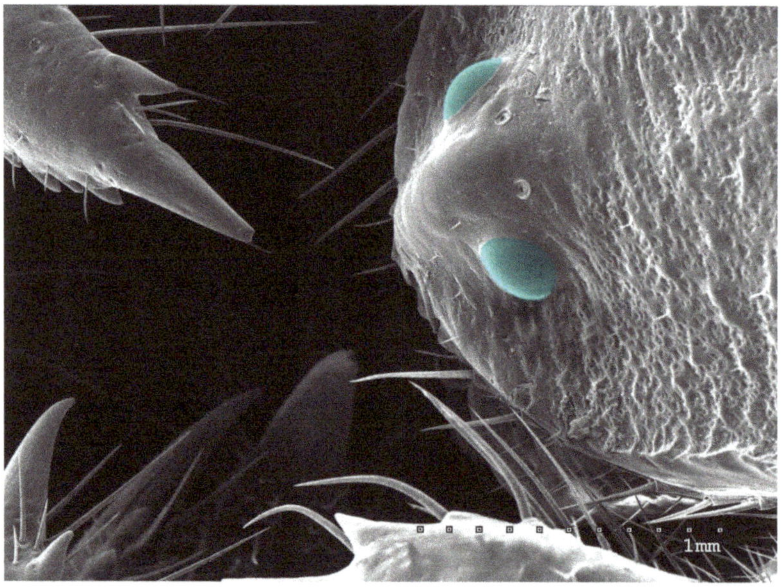

Median eyes.

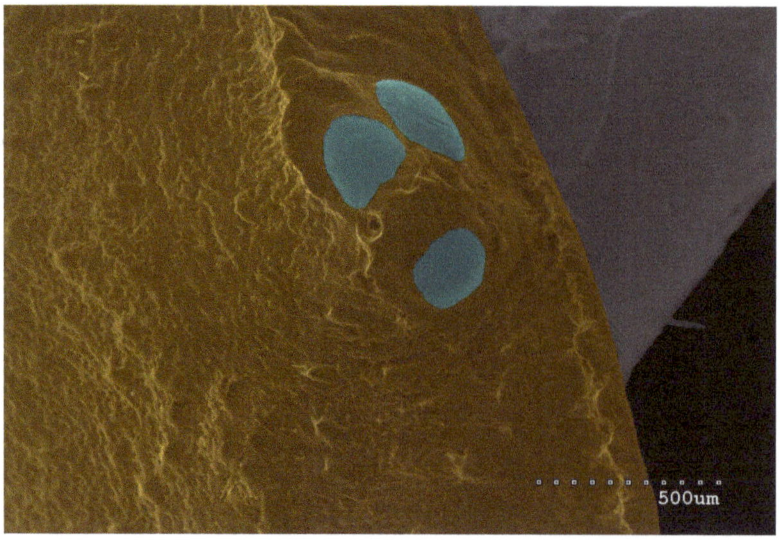

Lateral eye cluster of an early instar *T. stimpsoni.*

showed specimens could learn to run complex mazes as well as learn to avoid a mild shock by moving through compartments.

The *Mastigoproctus* exoskeleton is heavily outfitted with hairs, pits, and tubercles. © Michael Fanto

PROSOMA

The front section of the body is a combination of head and thorax also known as the cephalothorax. The mouthparts and legs are attached here. The surface is covered with a thick, elongate, pentagonal carapace in which the eight to twelve simple eyes are placed. Near the forward point of the pentagon lies a large pair of median eyes situated on either side of a raised ridge or hump. The median eyes face forward and may offer rudimentary imagery which may explain why my *Typopeltis* male sprayed me twice in the eyes but not once on the nose, cheek, or mouth. Much further back on each side are lateral eye clusters that normally contain three ocelli, two forward facing and one backward facing. The cluster as a whole is located on the folded down carapace edge and faces outwards rather than up. Each cluster can contain as many as five ocelli (also the maximum for scorpion lateral eye groups) but only on the small Philippine *Glyptogluteus augustus*, while the large *Thelyphonoides panayensis* has two small tubercles in the same position between the three eyes (Krehenwinkel et al. 2009). The various species said to have only three eyes per cluster possess what resembles a fourth eye in the middle. This structure shows up dark, nearly black, like the median eyes and is easily observed immediately following the molt (the three surrounding ocelli show up clear). However, the dark color is ocular pigment where the lenses direct images. The pigment is located beneath the cuticle and is not shed during the molt.

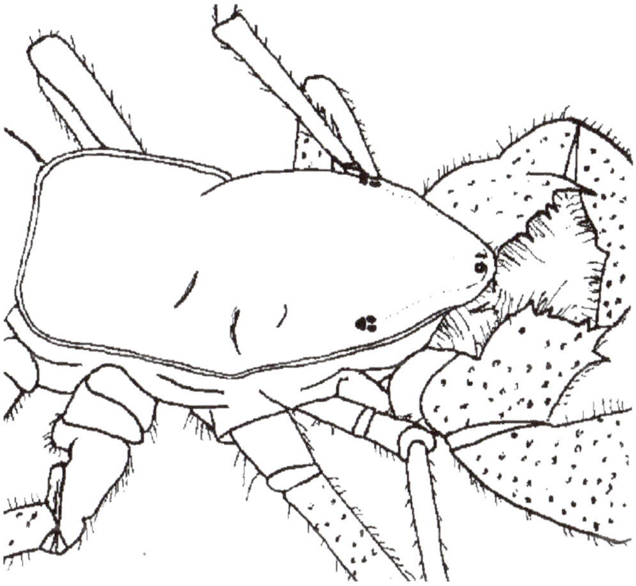

Eye placement on the carapace of *M. giganteus*.

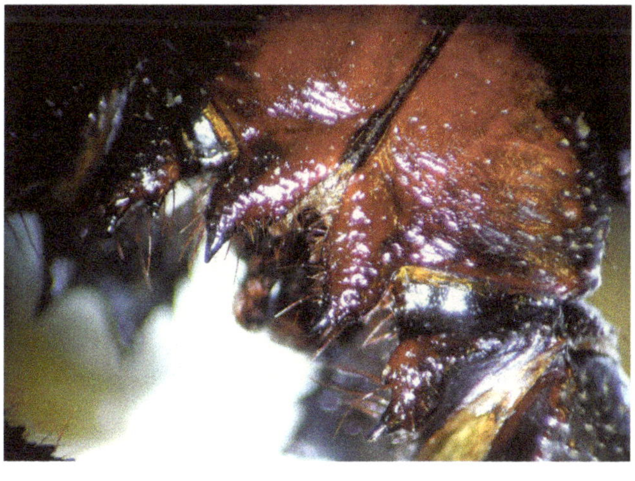

The preoral cavity is formed by the fused palpal coxae.

Lateral eye cluster. *T. stimpsoni*.

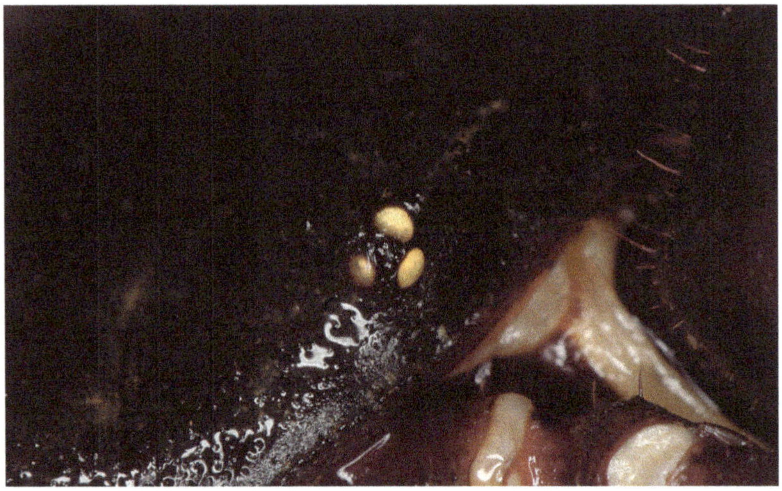

Lateral eye cluster. *T. panayensis*.
© Christian Schwarz

In *Mastigoproctus* and some other genera, the median eyes are located on either side of a slightly raised hump.

CHELICERAE

The chelicerae are the smaller pair of mouthparts that chew up the prey while the massive palps hold it in place. They do not contain silk glands as in pseudoscorpions nor venom glands like spiders. Whipscorpion chelicerae are similar in form to those of tarantulas and whipspiders in that they point downward and are composed of only two segments. The chelicerae are never greatly enlarged as they are on some spiders and whipspiders, and most palpigrades, schizomids, pseudoscorpions, and solpugids. The large basal segment usually has a few teeth but they cannot be seen without cutting away the thick mat of setae. Of course it is difficult to see much on a live specimen. The thick mat of setae ensures animals only ingest liquid particles and aids in grooming. Setae on the basal segment are long and smooth. The brush of "hairs" on the distal segment, or fang, are curved downward. Cheliceral setae must also be good conductors of sensory information as they lick the prey's exoskeleton clean rather than trying to chew it up. Other predators like whipspiders and mantids try to chew up hard prey and may end up giving up or missing much of the edible portion.

The chelicerae can be used to groom the antenniform leg and palpal tarsi. The palps help pull the leg tarsi in for cleaning as well as the tail. The abdomen is arched forward over the back in order to groom the flagellum. Unlike whipspiders whose fastidious grooming behavior is commonly seen, it is rare to see a whipscorpion grooming. Also this behavior can be restricted to nighttime so it is necessary to

Thelyphonus, *Typopeltis*, and some other genera have a narrow, raised ridge between the eyes.

The *M. giganteus* face.

make it a point to watch carefully for this behavior or it will never be seen.

PALPS

The pedipalps are thick and sturdy and built for burrowing and catching prey. They give the whipscorpions their characteristic look and can be sexually dimorphic. When not in use they articulate inward and look almost like boxing gloves. Palps rest and operate almost entirely within a horizontal plane, unlike those of schizomids, which rest in the vertical, or spiders and amblypygids that function in both vertical and horizontal planes. The palp shape does vary slightly between species and much more by gender, but across the entire order the variation is relatively low compared to scorpions or whipspiders. This conservation of form may result from whipscorpions all being burrowing species.

Palp segmentation is similar to the legs but lacks a patella and the tarsus has only two segments. The palpal coxae are huge and combine to form the bottom of the preoral cavity. The smaller trochanter is followed by a large femur and tibia, terminating in the basitarsus and distitarsus which form the claw.

Palps are used for feeding and fighting, which often amounts to the same thing. Even the fastest whipscorpions are not interested in chasing down moths or flies, so the thick, massive palps are perfect for tearing into heavily-armored, slow-moving prey like beetles and scorpions that many other predators ignore unless starving.

Despite being strong enough to tear through a heavily armored giant darkling beetle or the pelt of a small rodent, they rarely bite things that are not recognized as prey. This is somewhat

Large thick mats of setae obscure much of the chelicerae.

The palps of whipscorpions end in a pincer like scorpions and pseudoscorpions.

Claws of the chelicerae and coxae form the tearing surfaces in early instar *T. stimpsoni*.

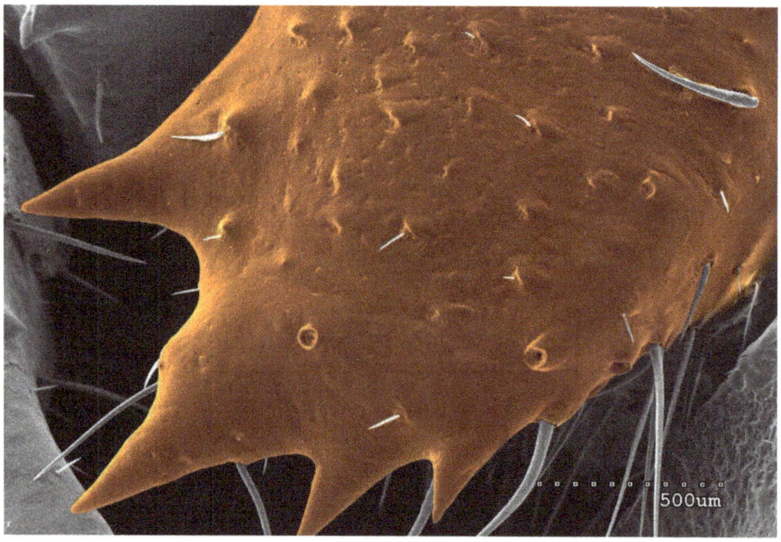

Spines on the dorsal surface of the trochanter can be different by species and even gender. Early instar *T. stimpsoni*.

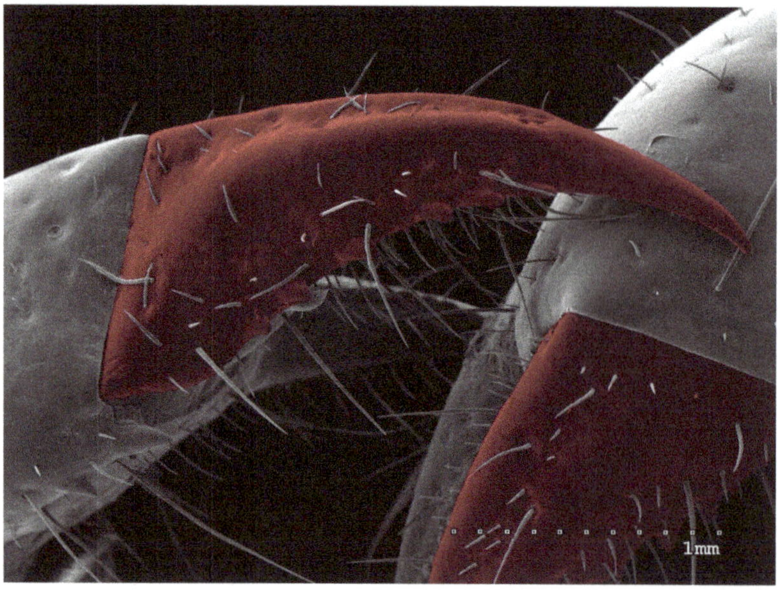

Palp distitarsi marked in red. Early instar *T. stimpsoni*.

Chelicera inner surface; the fang, articulation, and tooth structure are hidden by the extensive setae.

Chelicera outer surface with fang raised.

unusual since the raptorial catching mechanisms (whether palps, chelicerae, or legs) of various other organisms (amblypygids, tarantulas, crayfish, praying mantids, etc.) are used in defense and bite human fingers with almost reckless abandon whether or not it actually hurts. This behavior is not entirely unusual since Bess beetles, *Odontotaenius disjunctus*, have very strong jaws and almost never bite in defense. It is, however, strange the palps are held in a threatening posture for the deimatic display and still are rarely employed. It would seem the powerful defensive secretion has curtailed this defensive action.

Mating is another primary function of the palps discussed more thoroughly in the reproduction chapter. The large spine or finger on the segment behind the claw is commonly called the tibial apophysis. The apophysis is often used by the male to press sperm from the spermatophore into the female's genital operculum and so it is generally much longer or differently shaped in the male.

Palpal segmentation followed here is the original standard but there has been a recent move to call the tibia the patella and change the basitarsus into the tibia and consider the opposable finger of the claw the tarsus. This only matters in that the tibial apophysis is sometimes labeled a patellar apophysis, but it is the same thing.

Palp structure.

WHIPS

The antenniform front legs are known as the whips. They are highly modified with an incredibly long patella, cylindrical tarsal segments covered in an impressive array of sensory setae, claw pairs reduced to a microscopic sensory apparatus, and trichobothrial formation that differs from the ambulatory legs. As

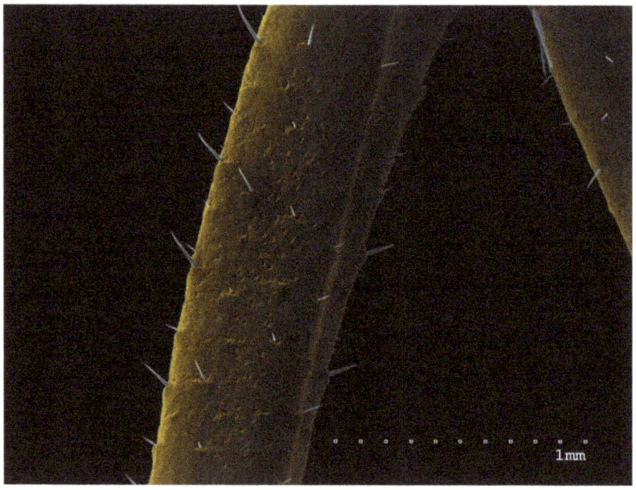

Antenniform leg tibia flattens out near the patella (femur at right). *Typopeltis stimpsoni*.

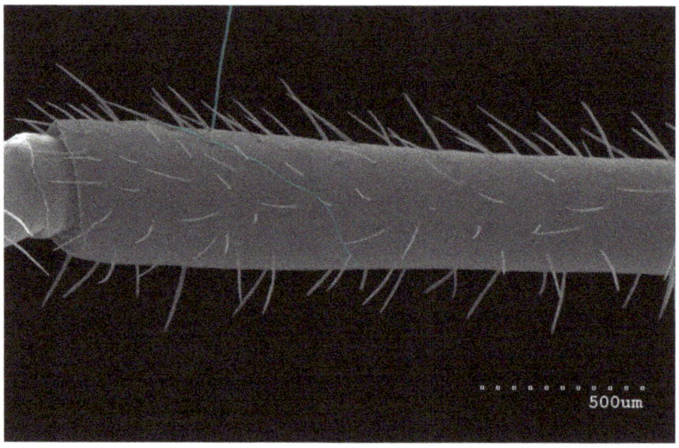

Antenniform leg tibia is rounded near tarsomeres (ventral surface of tibia shown).

Close-up of setae on antenniform leg tarsomeres.

with most arachnid legs, the segmentation is coxa, trochanter, femur, patella, tibia, and tarsus. The tarsus is divided into eight or so articles called tarsomeres. The first tarsomere is known as the basitarsus while the last is the distitarsus. The antenniform leg pair is thinner and stretches nearly twice the span of the walking legs. Though highly modified they are neither incredibly thin nor extensively segmented (as opposed to whipspiders) and the greatest span is only about 125 mm (5″).

In most species the whip tarsus has eight tarsomeres, *M. giganteus*.

The whips are highly articulate and are far more functional than a set of insect antennae. They can be folded around objects to sense the other side and to check out the contents of holes and cracks. During rest the whips are folded up at the joints so they do not stick out when the animal is hiding in a small crevice. During walking they are held out to probe the surroundings and can function independently (like chameleon eyes). The whips are usually held up and forward in the air during flight responses and forward in an arch for the deimatic display.

The whips are not just used to sense the ground and prey but are robust and shoved into holes to scare out prey. They are no more easily damaged than other leg pairs and show no

Sensory whip tarsomeres, shape and sensilla. *T. stimpsoni*.

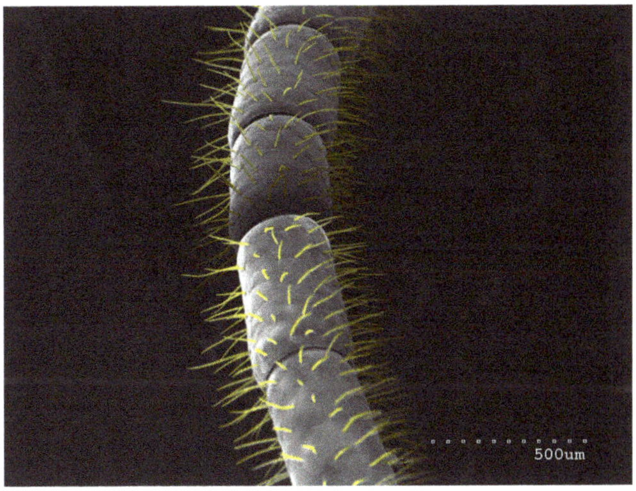

Whip tarsomeres are rounded and covered in setae.

propensity for being shed as a result of stress or trauma. (Opiliones and amblypygids rapidly autotomize sensory legs to specific points (McMonigle 2013)). Wild animals sometimes have damage to the whips but it can be between

The antenniform leg tibia is outfitted with a pair of trichobothria (view from one side). Early instar *T. stimpsoni*. The pit is nearly invisible to the naked eye while the filament is long, relatively straight, and evenly thin along its length.

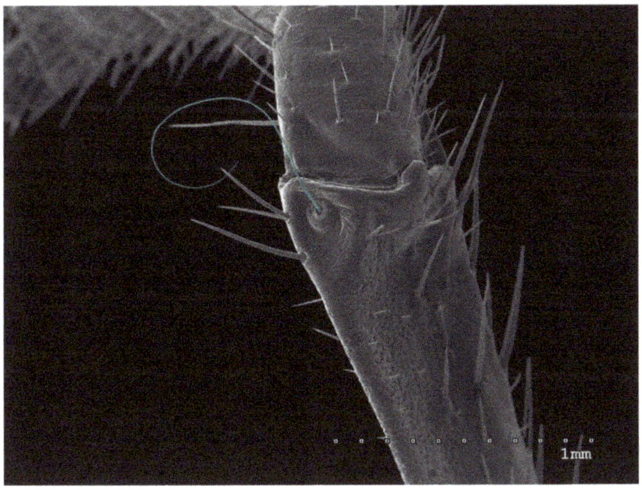

The unusual dome trichobothria is located only on the walking legs near the end of the tibia just before the first tarsomere.

or on any segment. I have never seen an animal missing both whips. Such a creature should do fine in captivity but could not survive long in the wild. The whips are very important for finding shelter, hunting prey, and for the female being dragged along during mating.

In this order the tail (flagellum) is sometimes confused and called a whip by the unfamiliar. If this were true the schizomids would be called short-whipped-whipscorpions, and

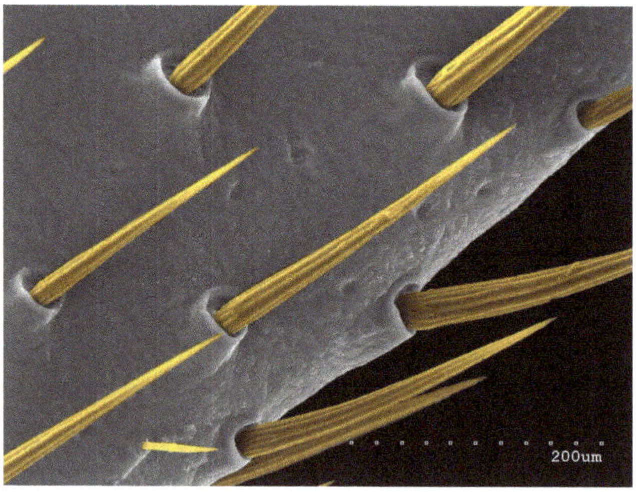

Setae on tibia of walking leg up close. Early instar *T. stimpsoni.*

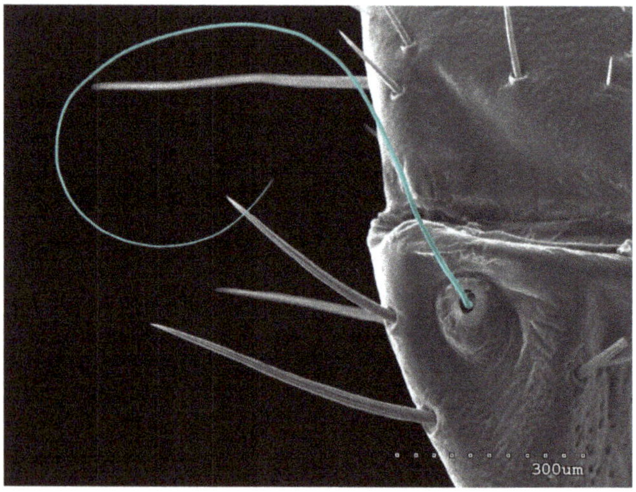

The large dome structure surrounding the walking leg trichobothrium is unusually large and visible with the naked eye, even on small animals. Also unusual, the filament is curled, thickened at the base, and thinning out along its length.

the amblypygids would also be known as whipless-whipscorpions. Other arachnids (beyond those from the four orders with whip in their common name) can have thin or elongate sensory legs, but they are short and touch the ground while walking or are uniquely adapted species from groups not primarily known for this feature.

WALKING LEGS

The ambulatory legs are built for forward locomotion and hold the body slightly above the substrate for walking. Limited backwards or sideways movement is possible but not very effective; if they need to run, they turn around. Although similar in segmentation to the whips, the walking legs have a short patella like the average arachnid leg. The patella is usually 25-35% of the length of the patella on the whips. (Example: *Typopeltis stimpsoni* late instar 2nd leg patella = 3 mm, whip patella = 12 mm.) The

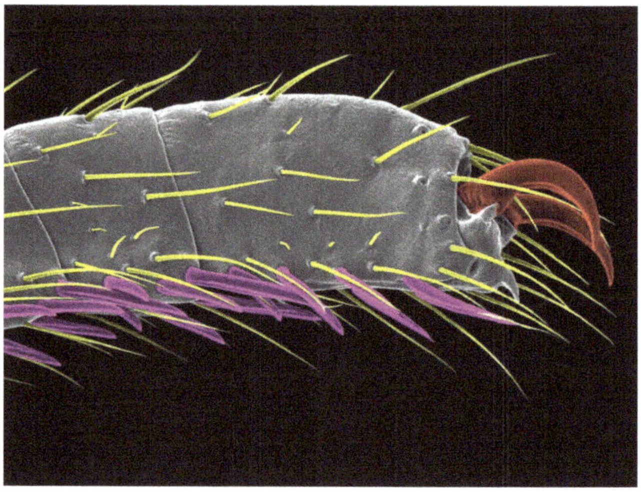

The walking leg tarsomeres have less tactile setae and large claws. There can also be large, thickened setae on the bottom surface. Early instar *T. stimpsoni.*

rear pair of legs is usually a little oversized and possesses a longer patella and tarsus, but they are still notably shorter than those on the whips. The enlarged rear legs may be used to lift the body higher to hold the broodsac above the substrate or balance the heavy opisthosoma during the threat display. The tarsal articles, normally four (compare to the whips at eight), are semicircular to square in cross-section. Depending on the species, tarsi can be outfitted with limited hairs or possess mats of thick setae towards the sides, combined with large flattened setae on the underside. There are no

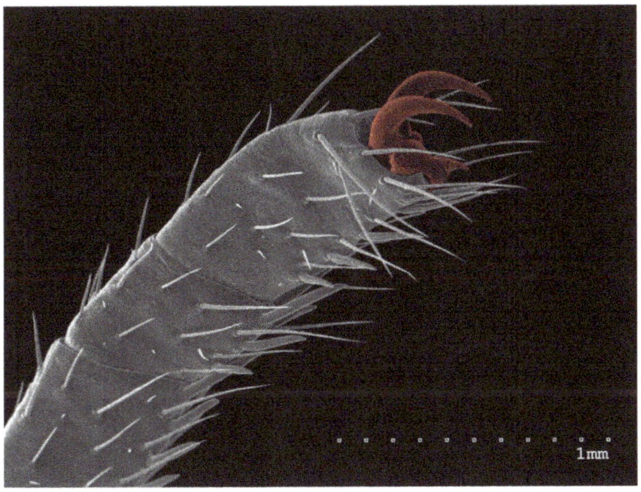

Claws of walking leg in red. *Typopeltis stimpsoni.*

Terminal whip tarsomere with microscopic claws marked red and tactile setae yellow. Early instar *T. stimpsoni.*

structures that would allow them to climb smooth vertical surfaces as found in most spiders and two of the five amblypygid families. Legs of species capable of rapid sprints such as *Typopeltis stimpsoni* are similar in structure to the legs of slower moving species like *Mastigoproctus giganteus*, but the tarsi are larger. Since *M. giganteus* are often found in

Notice the patella of each foreleg is many times longer than any walking leg patella.

There is a single, curled trichobothria near the end of the walking leg tibia which rises from the center of a large dome that is easily visible with the naked eye.

clumpy, clay soils, the smaller, less hairy tarsi may have nothing to do with speed. Less ornate tarsi may be necessary to prevent build up of large clumps of soil on the legs that lead to exhaustion and death. *Thelyphonus caudatus* are moderately fast moving and have somewhat plain tarsi, which may relate to native soil qualities.

TRICHOBOTHRIA

This sensory organ is found in most arachnids and its specific placement and quantity can be used to determine certain families, relationships, or species, because the placement and number are nearly invariable for a species. The complexity of structure explains why the number does not change; an extra trichobothrium would be almost as strange as an extra eye or antenna. Under a stereo-microscope the trichobothrium appears to be a simple hair (seta) set in a large socket, but scanning electron micrographs show it can be an extremely fine filament, often 2-3 microns in diameter and 2,000 microns in length (McMonigle 2013) attached to a tympanum-type membrane within a recessed pit. The trichobothria are primarily used for

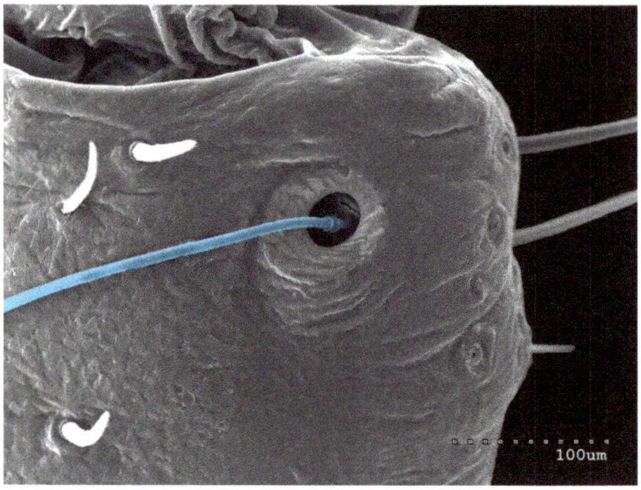

Trichobothrium close-up of pit and membrane (antenniform leg).

detecting airborne vibrations and air movement, though long filaments should be sensitive enough to detect sound. Santer and Hebets (2007) report that even the short trichobothria on the patella of amblypygids responds to specific sounds.

Whipscorpions only have a single trichobothrium located on the outer surface of the distal end of each walking leg tibia and one or two very long filaments on the tibia of the antenniform legs (compare this to amblypygids

On the whip there are two rather long filamentous trichobothria at the end of each tiba. There is no dome structure.

For most arachnids, each trichobothrial filament is attached to a microscopic membrane within a recessed pit. Walking leg of *Heterophrynus batesii*. This is similar to the trichobothria on the whipscorpion whips, not those on the walking leg.

that have about twenty long trichobothria on every walking leg tibia and a greatly reduced number of short filaments on the same area of the antenniform leg tibia). Trichobothria on the whips are of standard morphology. The whipscorpion walking leg trichobothria are different from those on other arachnids because they are each encompassed by a massive dome and the long filaments are particularly thick at the base and curled in shape. While there are around two dozen trichobothria on the patella and tibia of a single amblypygid walking leg (McMonigle 2013), and four dozen on a single palp of the scorpion *Heterometrus fulvipes*, the entire body of a whipscorpion has ten (Geethbali and Moro 1988). Scorpion trichobothria are located only on the pedipalps. Whipscorpion trichobothria are not used for identification because they rarely vary by genus, and when they do it is only on the whips and at most by a single filament.

PRE-OPISTHOSOMA

The opisthosoma (abdomen) is attached to the prosoma by a narrow waist which is also the first segment. From above it almost seems like a widened joint and there is a small tergite that spiders and amblypygids do not have. From below the narrow waist shows up as a tiny triangular first segment. The large second sternite shields the genitals and is known as the genital operculum. This and the following segment both have a pair of book lungs. Together the first nine segments are known as the preabdomen or mesosoma, as in scorpions. These segments have hardened dorsal plates known as tergites, and ventral plates called sternites, which are connected by the elastic and collapsible pleural membrane. Each plate-like mesosomal tergite and sternite has a pair of large indents, known as sigilla, related to muscular attachment.

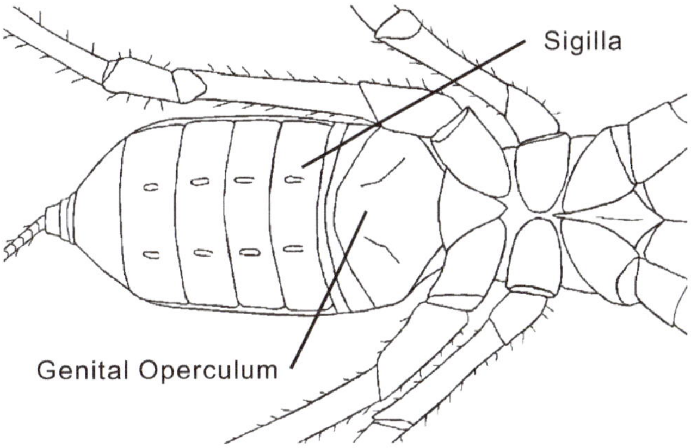

PYGIDIUM AND FLAGELLUM

The last three abdominal segments are greatly reduced and different. In these segments the tergites and sternites are fused together into a solid ring. Together they are known as the pygidium (postabdomen or metasoma). Like the fused, circular segments that make up the scorpion's metasoma, it is highly articulate with a higher range of movement than the legs, because it is used to direct the defensive chemicals. An approaching threat can be targeted through nearly 360 degrees around the body as well as straight up, though they tend

Pygidium and flagellum.

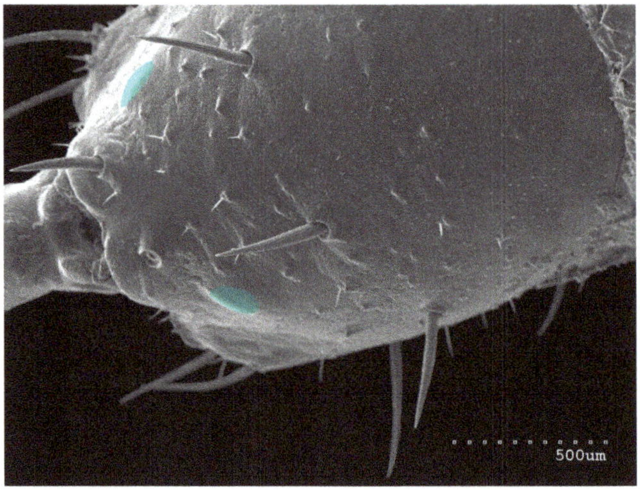

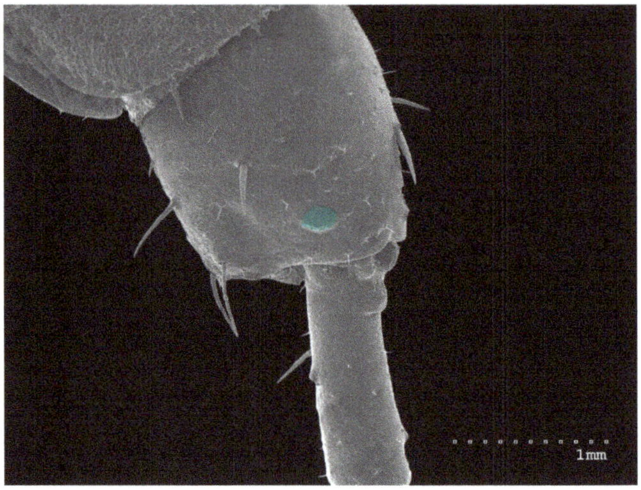

Ommatoid on the last segment of the pygidium (top and bottom). Early instar *T. stimpsoni*.

to turn around rather than spray backwards. Like scorpions, the terminal metasomal segment contains the anus, but in whipscorpions there are internal pygidial glands that produce the defensive secretion and pores on either side of the frasshole through which the defensive acid spray is emitted. There is also a pair of ommatoids on the last segment. Ommatoids are small, pale spots that reflect light at certain angles. They were originally thought to be light receptors since they resemble tiny lenses, but their function is unproven (Rowland and Cooke 1973). If the ommatoids are shaded and uncovered repeatedly on *T. stimpsoni,* the animals often exhibit flight responses, but they may simply be responding to slight air movement. The specific structure or lack of ommatoids has been used as a common character in describing species (as well as determining genera until recently (Haupt 2009)).

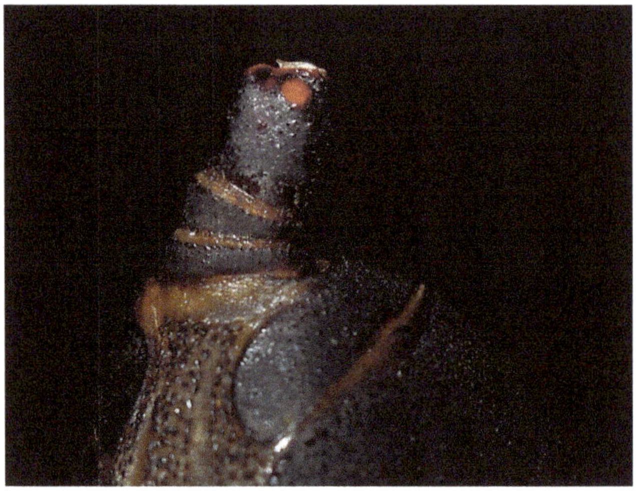

Ommatoid of an adult male *Typopeltis tarnanii*.

The name uropygi comes from the Greek and means *rump tail,* and probably refers to the narrow, tubular, tail-like pygidium (metasoma) and the whip-like flagellum (or telson) together as a tail. The flagellum is most commonly referred to as the whipscorpion tail, while in the scorpions the metasoma rather

Metasomal segments may retract in dead or emaciated specimens.

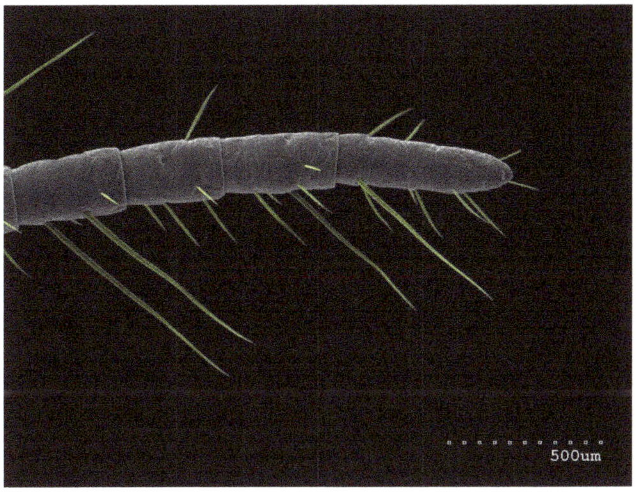

Setae are most numerous on the tip of the flagellum but still are sparse. Early instar *T. stimpsoni*.

than the telson is called the tail. Referring to either or both structures as a tail can be a little confusing, since the two structures are very different, and referring to either as half a tail is even less clear, since they are each composed of multiple segments.

The pygidium terminates with the flagellum, which is analogous to the scorpion's telson, but is terribly different in form and function. There is no stinger and no internal venom glands. Depending on species, the flagellum is usually composed of twenty to thirty articles with a maximum of fifty for *Thelyphonus* (*Tetrabalius*) sp. from Borneo. Although often considered sensory in function, the flagellum is relatively barren of a sensory apparatus. The only area of the body with less sensory setae than the flagellum is the upper surface of the carapace. The flagellum does not function like antennae, but more like the abdominal cerci of insects or uropods of isopods (cerciform, not antenniform). During hunting, walking, and probing, the flagellum is held backwards and upwards, away from any threat. The flagellum is never used to probe objects nor prey, and whipscorpions never walk backwards probing objects with their tail. Still, it is a rear guard and outfitted with limited setae that offer tactile information. Lack of extensive setae and actual function suggest it may be more important for directing the defensive response. Only in the threat display can it be rotated towards a threat by movement of the pygidium. Some people confuse movement in the threat display with sensory function. Animals missing the flagellum can still spray from the pygidial glands, but lose the ability to direct the defensive fluid with accuracy or distance. This is easily tested since many specimens lose the flagellum when

Notice the rather sparse setae (compared to most other body surfaces) on articles of the flagellum. *M. giganteus*.

Central area of a *T. stimpsoni* flagellum.

The end of the flagellum has the most setae but it is still relatively sparse. *T. stimpsoni*.

they are still alive. Tail-less whipscorpions not only lose accuracy but also are less likely to spray.

The strangest aspect of the flagellum is it seems to fall off with almost reckless abandon. Tail-less whipscorpions are incredibly common in captivity and are not rare in the wild. Some rare species are known only from specimens that are missing the flagellum and this appendage is undescribed for these (Rowland et al. 1973, Viquez 2003). Up to 10% of wild-caught specimens are missing the flagellum, while the percentage can be much higher in captivity. If the molting parameters are imperfect, loss of the flagellum during the molt is likely. I have seen a handful of captive-hatched 2nd instars drop the flagellum in shipping for no apparent reason. Later instars and adults rarely lose this appendage in shipping. Animals that have been kept in captivity for a few years often lose the tail months or years before they die of old age. When two animals fight, the loser almost always has the flagellum ripped off—if the whole animal was not eaten. Even when the tail is not missing a close inspection reveals many wild animals that seem to have a complete flagellum are missing part. The pygidium is never lost since it is part of the abdomen and would likely result in a rapid death. Flagellum loss is not related to autotomy since wild animals are found with tails broken off at nearly every point

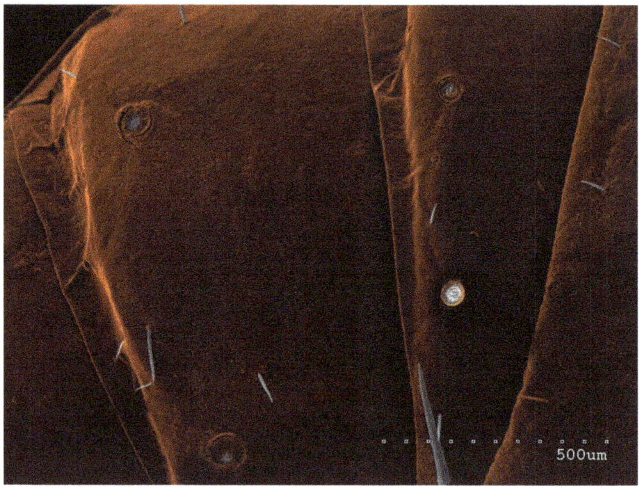

Pygidium segments fold into each other as they reduce in size. The folded connective area is pale colored but SEMs show texture. The rightmost segment is the last mesosomal tergite.

(whether this is mostly caused by natural damage or collecting and shipping method imperfections is uncertain). Autotomy in spiders, whipspiders, and various creatures involves one, rarely two, specific joints on an appendage. Easy loss of the tail is why uropygids should be preserved in dry mounts if aesthetics are a concern. Specimens preserved in liquid

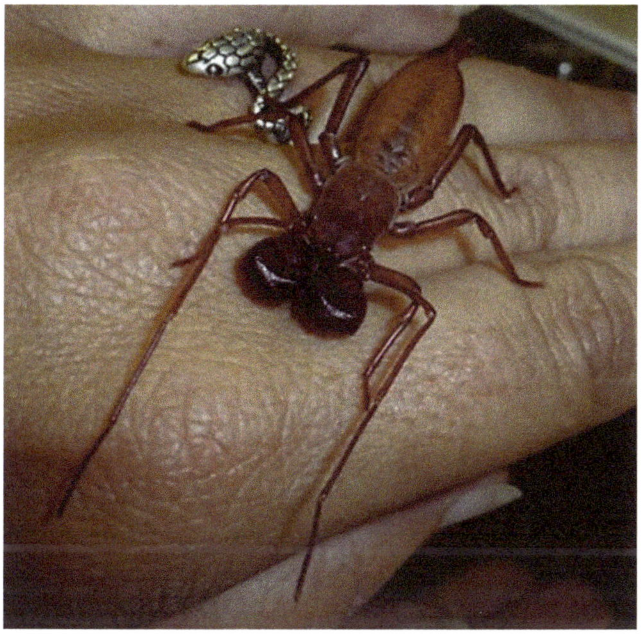

This rare, wild-caught red form of *M. giganteus* from El Paso was caught without a flagellum. © Traci Roach

commonly lose the flagellum, whereas dry mounted specimens do not.

When the whipscorpion is hiding in a shelter the pygidium can be rotated forward or to the side so the flagellum rests flat against the top of the body. Unlike the antenniform legs that are highly articulate, can fold up, or be used to probe the environment, the flagellum has no muscles and can only be moved using muscles to change the orientation of the mesosoma and metasoma. However, there are tendons within the tail that allow it to be straightened when bent.

CHEMICAL DEFENSES

Onlookers do not realize the only real defense these frightful looking monsters possess is a spray of tiny droplets which impart the delicious taste of vinegar potato chips. The primary active ingredient of the defensive spray is acetic acid, the active ingredient of vinegar. Such a defensive fluid seems pretty silly from a human perspective, since we employ this acid as a food flavoring. However, it is very effective against predators and the concentration is much stronger.

Vinegar used in food preparation is usually 4-8% acetic acid in 92-96% water. The strongest vinegars often are not legal to sell for food use and can be as concentrated as 30% acetic acid. The secretion of *M. giganteus* contains about 25% water, 10% octanoic acid, and 60%+ acetic acid. There are some trace related fatty acids as well. The exact composition varies by individual and the ratio of acids may differ significantly in some specimens (Eisner et al. 1961, Schmidt et al. 2000). Tests of the Asian *Typopeltis crucifer* have shown a similar chemical make-up (Yogi and Haupt 1977). The Japanese *Typopeltis stimpsoni* defensive fluid seems to fall just above the upper range of *M. giganteus* results with acetic acid at 78%, octanoic acid at 6%, and 16% water (Itokawa et al.1985). The results reported by different authors are more similar than they seem at first glance because some use the name caprylic acid and others use octanoic acid (the same compound). Furthermore, reports of around 84% acetic acid for *M. giganteus* are percent relative to components other than water (after the water has been removed from the equation). Keep in mind that percentages which do or do not include water in the equation must be translated with a calculation to make a comparison.

Most secretions have a significant secondary active ingredient that makes the acetic acid more effective by etching waterproof exoskeletons or dissolving water repellant oils in animal pelts. The most common, octanoic acid may make the acetic acid more effective since it solubilizes the epicuticular wax layer which protects many arthropod exoskeletons. Octanoic acid is an oily substance, barely soluble in water and so requires the highly concentrated acetic acid as a carrier. Hexyl acetate is a solvent with a fruity odor that mostly takes the place of octanoic acid in *Hypoctonus siamensis* (Haupt and Mueller 2004) and probably other Asian species according to smell. The secretions of an unidentified *Typopeltis* species from China has 2-nonanone and 2-octanone instead of octanoic acid (Haupt et al. 1988). These chemicals irritate the eyes, skin, and respiratory tract of humans when concentrated, and are nearly insoluble in water.

In addition to primary and secondary active ingredients, there can be related trace ingredients or minor components. Schmidt et al (1998) discovered trace amounts of 2-ethlyl-hexanoic acid, (Z)-5-octenoic acid, hepanoic acid, decanoic acid, hexanoic acid, and (E)-5-octenoic acid within the pygidial glands of *M. giganteus*. It is unknown if trace ingredients are accidental byproducts or functional components capable of offering microbial protection or offensive improvement at extremely low concentrations.

In biological organisms, acetic acid is normally produced by converting sugar into ethanol and then oxidizing ethanol to create water and acetic acid, although some anaerobic bacteria can transform sugar directly into acetic acid. Vinegaroon pygidial glands must use enzymes to process the sugars into acetic acid, which solubilizes the lipophyllic octanoic acid and other longer chained components. Bacteria cannot be used in the glands because the environment within the pygidial glands would kill them. There are many acetogenic bacteria but those that directly catalyze sugars into acid die when the percent exceeds 1% and even the most virile bacteria that require a longer time because they change sugar into ethanol first,

The flagellum can be folded forward through articulation of the pygidium.

and then to acid, die at 20% (Sim et al. 2007). Not only are high concentrations of acetic acid antimicrobial, octanoic acid is commonly used to kill bacteria.

Although a few tests have been run in the last fifty years, there is much to learn about basic composition of different fluids and their synthesis. Present test results that show a similarity between *M. giganteus* and *T. stimpsoni* fluids seem to be in error. Although the defensive secretion of *M. giganteus* smells incredibly strong of vinegar, *T. stimpsoni*'s fluid is many times more pungent and burns the nostrils (I have tested this on a few friends as well). Also, skin contact with *M. giganteus* fluid from various Arizona and Texas specimens has resulted in no sensation. In stark opposition, *T. stimpsoni's* fluid causes a painful stinging sensation after about thirty seconds on skin located on the eyelids and the back or sides of the fingers and hands (no pain is usually felt on the thick skin of the fingertips or inner palm). If not washed off the stinging sensation continues for at least ten minutes. Differences in many species have not yet been tested. *Thelyphonus caudatus* smells like vinegar but also has a fruity component to the odor (probably due to hexyl acetate), while *Thelyphonus manilanus* secretion does not smell like vinegar, although the secondary ingredient may mask the acetic acid odor. A few popular texts employing limited research claim formic acid or chlorine can be a primary component (McGavin 2000, Hernandez 2001, O'Toole 2002) but neither chemical has ever been found in any secretions tested (Eisner et al. 1961, Yogi and Haupt 1977, Itokawa et al. 1988, Schmidt et al. 1998, Schmidt et al. 2000, Haupt and Mueller 2004), and the authors are forwarding a past rumor laid to rest (Breene 2000). The rumor seems to have started with Levi and Levi (1968) who claimed no lab test results but mentioned, "Other tropical species *smell* of formic acid or of chlorine."

Not only does the ratio of components vary somewhat according to the testing methodology, species, and individual, but it can vary also within a single specimen. An animal that has discharged multiple times will require a few days to recharge, but the fluid can still smell

TABLE 2. ACTIVE INGREDIENTS

	Acetic acid	Octanoic acid	Hexyl acetate	2-Nonanone	2-Octanone
Alternative Name	Ethanoic Acid	Caprylic Acid	Capryl Acetate	Heptyl Methyl Ketone	Hexyl Methyl Ketone
Molecule Components	$C_2H_4O_2$	$C_8H_{16}O_2$	$C_8H_{16}O_2$	$C_9H_{18}O$	$C_8H_{16}O$
Common Uses	Food flavoring, PVC and glue production, buffers, preservatives	Industrial antimicrobial pesticide	Solvent for fats and polymers; fruity food flavoring.	Food flavoring or fragrance	Food flavoring or fragrance
Solubility in Water	Fully	Minimally	Somewhat	Minimally	Minimally

notably weaker. The quantity and propensity to spray can also vary within the same individual. Most species commonly do not spray if allowed to walk onto a hand, rather than being grabbed and placed on a hand or harassed. Quantity is often greater with rough handling and less with constant handling. Vinegaroons discontinue use of the spray with frequent handling as a result of becoming used to the stimulus, not just because they run out of defensive fluid temporarily. Lastly, recently acquired wild specimens often spray stronger smelling fluid, more often, in greater quantity, and a further distance. Specimens in captivity seem to "relax" after a while and never seem to be as virile in defense even when they have not been handled for months. In general, the volume of defensive fluid is equivalent to the size of the animal. The 2nd instars are capable of spraying a smelly and impressive defense, but nothing like they are capable of at full size. All the above differences relate primarily to quantity or external differences; the chemical composition of the spray is relatively constant whether the individual is male, female, or immature (Schmidt et al. 1998).

The powerful odor is very impressive for the fluid volume. Even the largest, most productive vinegaroon rarely excretes more defensive fluid than would be contained in a fraction of a droplet from a common eyedropper. A huge bottle of vinegar can be quite smelly but less than a drop is relatively odorless. I have compared the defensive secretion odor with acetate buffer (46% acetic acid, 15% sodium acetate, balance water) from my lab and even this smells relatively unimpressive at the same volume (and not even remotely on par with *T. stimpsoni* secretion).

Sometimes the spray can be flung more than a foot, but usually it barely shoots a few inches. Reports of long distances over two and a half feet for *M. giganteus* (Martin and Stoops 1995) sound like exaggeration, even if they resulted from placing specimens on a large sheet of litmus paper and measuring the furthest tiny overspray droplet of an especially powerful individual, rather than the effective distance, and certainly not upward projection. The 20″ (50 cm) of Bruins (1999) is a more realistic maximum. *Typopeltis stimpsoni* is a far more effective shooter than *M. giganteus* and it is hard to imagine even tiny, vertical, overspray droplets exceeding two feet. The cloud of tiny droplets is clearly visible even in small specimens of this species and large immatures only project about 2.5″ (6 cm) upwards—only the adults can greatly exceed that. Distance is not terribly important since it will not be wasted on a distant threat. Defensive fluid only needs to get in the face or eyes of an animal already close enough to eat it. Humans can rinse the spray out of their eyes with water, while a small mammal may feel the stinging and discomfort for a long time. Usually they spray human hands when picked up, but twice I had an adult male *T. stimpsoni* shoot me in the eyes from a distance of about ten inches as I went to grab it (instead of directing its spray towards the menacing hand as usual).

I have heard stories of people catching a group of *M. giganteus*, placing them in a small container, and later finding them and every other creature placed within dead. I have kept specimens of a few different species, including developing juvenile *M. giganteus*, for years in rather small containers with only a few pinholes to prevent deaths caused by desiccation. Over the years they have sprayed numerous times when disturbed and yet not a single animal has become ill from its own defensive spray. Such reports may be result of overheating or suffocation from too many animals and lack of any ventilation holes rather than the

A

Molt series (A-C) from 2nd to 3rd instar, June 5-6, 2007. Molt to 2nd instar took place July 5, 2006. Captive-hatched *M. giganteus*. © Michael Fanto

B

C

Three-year-old *M. giganteus* a few days after a molt. © Michael Fanto

defensive fluid itself (Franklin 2002). According to Wade Harrell (pers. comm.) during field collecting he has seen individuals die in small vials with holes, while those in larger containers survived. He believes the cause may be they "gas themselves to death," especially during collection because they are more prone to spray in greater quantity when freshly caught. Though easy to avoid, the exact cause of collection fatality remains uncertain. Justin Schmidt (pers. comm.) reports he has ". . . sealed adult *M. giganteus* in 1 liter containers that are air/pressure tight (as measured for a week in a vacuum desiccator) and held them for over a week. They do not die from gassing or lack of oxygen."

MOLTING

Growing whipscorpions molt off the substrate surface like scorpions or tarantulas and likewise must molt in shelters in nature to survive. They do not need to hang like amblypygids, mantids, or web spiders. Wild *M. giganteus* tend to molt between March and May each year and this usually carries over to captivity. However, great delays are possible under captive situations and sometimes they will wait till the following spring. Tropical species can molt at any time. Unlike some arthropods that quit feeding weeks or months before the molt and move very little, whipscorpions move somewhat normally and may even feed within a few days of the molt. They pop the prosomal shield, pull out the legs and finally the flagellum. At first the new exoskeleton is very soft and white in color, but it darkens considerably within a few hours. Small specimens can take a week to obtain final coloration and maximum rigidity, while large specimens can take even longer. The exuvium is not eaten.

Small animals must be kept humid since dehydration will result in failure to molt properly and death. A certain amount of blood pressure is required for a successful molt and spraying an animal after it has begun to molt is practically useless. It is too late then to replenish internal fluid levels and the exoskeleton is not water-soluble (or the animals would melt when it rains). Some losses are experienced in molts that cannot be easily blamed on external factors. The animals simply do not make it fully out of the old exoskeleton and later die. The results resemble dehydration. These individuals may simply be weak or have genetic defects, but nutritional deficiency is a possible culprit.

Immature animals that have lost the flagellum can regrow it, albeit in shorter length, in a single molt. Some regeneration of damaged legs is also possible though not as extensive.

Since vinegaroons have thick body armor everywhere, from the extremities to the segmented abdomen, and they do not eat their molts, they are one of very few arachnids whose cast off exoskeleton can look as good as a dried specimen. The exuvium can even look better since it is not discolored from decomposition. It must be removed from the cage and dried as soon as possible. The hardened plates can sit in a damp cage for years without any degradation, but the membranes between the plates and segments can begin to rot away after less than a day of high humidity. Of course this is only for immature stages since adults do not molt.

LIFESPAN

Whipscorpions grow slowly. *Mastigoproctus giganteus* take one year for a single molt and must molt four times after dispersal to reach adulthood. Even though they are not forced to endure a period of hibernation, immatures still require a full year between each molt in captivity. Sometimes a specimen can take two years to get through a single molt. Animals with greatly delayed molts could take much longer to reach maturity, but greatly delayed specimens often end up dying or getting stuck in a later molt and never mature. Some of the tropical species might grow much faster but documented captive development to date shows a year between molts is the norm. Some Bornean *Thelyphonus* captive-hatched second instars that molted on September 9th, 2009, molted again from second to third instar, in August 2010. A captive-hatched male *Typopeltis* from Malaysia took five years to reach maturity (Cain Eyre pers. comm.). *Typopeltis stimpsoni* appear to grow no faster. Partial and complete life cycles have not been charted for other thelyphonids, but a faster-growing species, if it exists, would still be likely to require at least a few years from leaving the mother to sexual maturity.

Like most arachnids, whipscorpions do not molt after reaching adulthood though they survive longer after the ultimate molt than nearly any arachnid with a terminal molt. Since molts can run like clockwork according to seasonal hibernation in nature, the long adult lifespan may be what helps keep different year's broods from becoming genetically isolated. Wild-caught adults rarely live less than a year. Half a dozen wild-caught adults of the Malaysian *Thelyphonus caudatus* I have data on all lived two to three years before expiring. Five years is the longest I can certify a wild-caught adult *M. giganteus* has lived in captivity, but three years is the more common span. In the wild, *M. giganteus* likely live just as long. Marked adults have been recaptured the following summer (Schmidt 2003). All available data contrasts greatly with a rare report (Franklin 2002) of an *M. giganteus* female specimen living for nineteen years in captivity. Even if it had been acquired at 2nd instar and took five to six years to reach maturity (and then continued in maturity for another five years) such a claim would still be eight years longer than the maximum available data suggests could be possible.

PAST AND FUTURE

The seeming lack of variation in species found around the world may relate to minimal genetic drift since they grow slowly and reproduce in small numbers. Fossils look very similar to living whipscorpions. During the Carboniferous when cockroaches had long ovipositors, whipscorpions looked the same as they do today (Dunlop and Horrocks 1996, Roth 2002). It is very strange that all vinegaroons look similar, while other groups show a greater morphological diversity, but this happens also in other groups, like the horseshoe crab and tardigrades—all the species in the world look the same and their fossils do as well. This might be due to a low genetic plasticity and a very specific niche they live in. The reason therefore is not well understood. Similar with pigeons, dogs, or goldfish, we manage to develop all these very different breeds from variations existing within their genetic code, while horses do not show this strong diversity. Whipscorpions are primarily tropical, so some arachnologists believe the presence of *M. giganteus* in subtropical North America may indicate relict populations left over from the Pleistocene (Franklin 2002). However, during the Pleistocene it was colder in North America and it is unlikely that these species could occur in an even colder climate. *Mastigoproctus giganteus* shows some adaptations to dry climate by avoiding desiccation during the day in deep burrows, being able to extract humidity from sand, and by the greater size (much better volume to surface ratio). This is clearly a species which moved outside the usual climate zone.

In Pleistocene times, cold periods interchanged with warm periods, creating an oscillation of animals and plants being driven south during glacial periods and re-conquering lost land during warm periods from refugia. One such refuge was southeastern North America, as evidenced by Tertiary relics like magnolia, swamp pine, alligators and other creatures (in mantids e.g. *Mantoida maya*). These taxa are Tertiary rather than Pleistocene relics. They have survived in refugia while their populations farther north were killed off by glaciations and cold temperatures. On the other hand, during cold periods sea level is lower, facilitating or promoting dispersal from the Caribbean. Getting back to *Mastigoproctus*, it is likely that populations from Florida are a Tertiary relic of a once contiguous distribution in southern North America, while those from the southwest have re-invaded lost ground, but this needs confirmation. Considering the Carboniferous or so age of the order, its homogenous appearance, and slow reproduction, some ten thousands to millions of years of isolation may not be enough to create much of a morphological difference.

Whipscorpions are often found near water sources, such as rivers, especially in dry forest areas. Worldwide most of them are found in the tropics in very wet climate; only a few are in subtropical and temperate climates, and these are surely found close to water. Waterfront property is often valuable real estate for humans. *Mastigoproctus giganteus* is a widespread and relatively common species, but it has become locally extinct where humans have built shopping malls, industrial complexes, and new housing developments. However, populations seem to have popped up in areas where man-made structures such as in-ground pools, spillways, golf courses, and water meter boxes are found (though they may simply be easier to find with increased activity due to greater water availability). Possibly the greatest threat in the long run (assuming powerful residual pesticide use continues to be restricted) is continued building of roads and increased traffic. Roads cut off populations and serve as effective squashing devices for those in search of mates or new territory. Nevertheless, it seems sidewalks and roads in New Mexico and Texas may create new habitat as populations are often found beneath them (Breene 2000, Lorbiecke pers. comm.). Isolated populations in Florida seem to wander or disperse very little, but whether this is related to wandering behavior being literally squashed out of the population by the incredible traffic volume in the last hundred years or habitat isolation over a much longer time is not known.

Male *M. giganteus*. © Michael Fanto

Female *M. giganteus*. © Michael Fanto

Sexual dimorphism of *M. giganteus* palps (male at left).

3
REPRODUCTION

THE FIRST STEP in mating or breeding is identifying the male and female since no known vinegaroon species are capable of parthenogenesis (just some of the short-tailed whipscorpions). Males and females are sexually dimorphic at maturity, but small animals appear to have female features and even subadults are difficult to sex. On maturity the adults display a number of differences that vary by species and genus, much less so by individual. Differences tend to be obvious when looked for. If the specimens do not display any of the following dimorphisms they are the same gender or immature.

Palps of an adult female Chinese vinegaroon that had lived two years in captivity.

The tibial apophysis of a female *T. stimpsoni* has a normal, tapered shape and ends in a point.

Male palps are longer, often larger than in a female of the same size, and the segments are more angular than rounded in appearance. I think the palps of the mature *M. giganteus* male looks nothing like the female's, and could tell them apart from ten feet, but a friend told me he could not see the difference on close-up photos.

Notable spines on the dorsal trochanter of the pedipalp may be fewer, smaller, or missing on mature males depending on the species.

The tibial apophysis (often the largest finger or spine on the palp) is longer on adult males; furthermore in the genera *Hypoctonus, Labochirus*, and *Typopeltis* the tip is unusually shaped and spiny (Rowland and Cooke 1973) or looks like a short corkscrew.

Genital operculums of a male *M. giganteus*.

Adult male *T. stimpsoni* genital operculum.

Female *M. giganteus* genital operculum can be difficult to discern from the male`s.

Even among the strongly sexually dimorphic species, the genital operculum of the immature generally does not betray the identity of its gender. Even this 5th instar (subadult) *T. stimpsoni* appears decidedly undecided.

Genital operculum of a male *M. giganteus*.

Typopeltis tarnanii adult male genital operculum.

Typopeltis tarnanii adult female genital operculum.

Genital operculum of an adult female Chinese vinegaroon.

(Top and Bottom) Modified tibial apophysis shape of the male *T. stimpsoni* from different angles.

Adult male *T. tarnanii* with folded palps.

The trochanter has lost its spines, basitarsal finger is reduced, and the tibial apophysis is enlarged and terminates in a process that looks like a short corckscrew or hammerthumb. *T. tarnanii*.

The adult female`s palpal trochanter is spiny, the finger on the basitarsus is larger, and the tibial apophysis is normal. Immatures of both sexes have the same morphology. *T. tarnanii*.

The whip tarsomeres of this *Typopeltis tarnanii* adult female do not look notably different from a male`s.

In some species males have a notably shorter flagellum, though this could easily be confused with damaged specimens.

Female *Typopeltis*, *Thelyphonus*, and some other genera have one or more spines on the whip tarsomeres, but this is extremely hard to see without magnification.

In most species the large plate under the opisthosoma known as the genital operculum looks very different for the male and female. The male operculum is often much larger and does not have a small groove at the end. While most species are very easy to sex this way, the genital operculum of male *M. giganteus* is variable and the difference in shape more subtle.

The elastic pleural membrane on the sides of the abdomen can be notably wider, or at least stretched out, on the adult female.

Adults can be placed together during the day and their reactions should be easy to read unless the response is indifference. If adults are unreceptive (or the same sex), and close attention is not paid, one will quickly chop off the other's tail; much longer and one will end up being eaten. Receptive adults can remain together for many hours, possibly days, after mating has ceased, but it is safest to separate them as soon as the dance has been completed. Unlike spiders and mantids, there is little to no chance of the animals eating each other during the mating steps.

Mating involves a dance that is more physical than the average amblypygid ritual, and more time intensive than the scorpion dance.

The male *Thelyphonus caudatus* has a naturally shorter flagellum.

The *M. giganteus* dance consists of specific phases that can last a few hours each. The total time required is usually eight to twelve hours. First, they face each other and the male takes hold of the female's whips in his chelicerae. Next, he positions her behind him and pulls her along as he looks for a good spot to make a spermatophore—usually a flat piece of bark or rock is preferred over loose substrate. Formation of this structure takes a few hours. When the spermatophore has been formed and taken into her genital opening, he grasps her with his pedipalps and pushes the sperm into

Stages 1, 2, 3 of *M. giganteus* courtship.
© Michael Fanto

Early courtship for *Thelyphonus caudatus* in order. *Thelyphonus* are not known to have a stage 3 where the male grasps the female with his palps to aid sperm insertion of spermatophore. © Michael Fanto her. A broken spermatophore is a good sign of success. Other species have similar dances, but *Thelyphonus* males, for example, do not use the palps to press in the spermatophore (Weygolt 1970). Some of the gender differences in whip structure and tibial apophysis shape are easily related to parts of the mating process.

After mating, females retain developing eggs in the opisthosoma for a number of months before extruding them inside a membrane. A successful mating does not necessarily result in reproduction and even wild-caught females that mated in the summer or fall may wait till the following year before mating and trying again. Shane Burchfield (pers. comm.) reports that out of a hundred or so mature females collected, only five or ten end up forming a broodsac each year. In reproductive females, the eggs develop on the ovaries until they reach full size. Probably the largest eggs are those of *M. giganteus,* with a diameter of 3.5 to 4 mm. The ovaries take up most of the mesosoma, and the eggs grow like grapes along the entire ventral surface just behind the sternites. In some species the fully developed eggs can be seen through the intersegmental membranes between the abdominal sternites a few days or weeks before the broodsac is extruded.

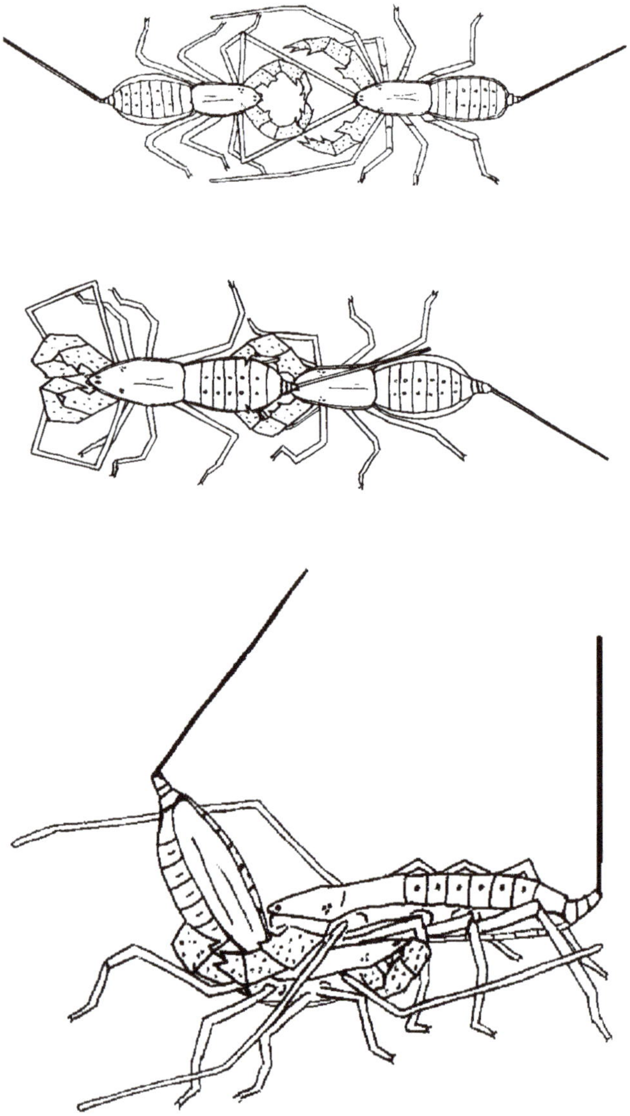

Stages 1, 2, 3 of courtship for *M. giganteus*.

Used spermatophore remnants. *M. giganteus.*

If fed, the female will continue to eat until the very day before the broodsac is extruded. However, *M. giganteus* females are known to dig out an underground chamber around September and not surface with their 2nd instar immatures until June or July. Females are unable to capture food when sealed in the chamber. It seems unlikely tropical species

Broodsac early formation.

Freshly completed broodsac.

The broodsac membrane wrinkles and contracts against the eggs as the female withdraws fluid over twenty-four hours. Most of the fluid is withdrawn over the first few days.

from humid forests could go without food for nearly so long.

The broodsac is extruded in one piece as a gelatinous membrane containing eggs and fluid. The eggs never touch the air and the base remains firmly attached below the genital operculum. Some females lie on their backs during broodsac extrusion while others stand upright. The broodsac shape is affected, but it is uncertain if the behavior has a bearing on the success of gestation. The membrane is very soft and sticky at first, so pieces of dirt will stick to it if the female is forced to move around. The fresh broodsac hangs like a bag of water because it is filled with fluid that was used to push the eggs out into the membrane. For the first day or two the eggs are free and move inside the sac if jarred. Most of the fluid is withdrawn back into the female's body over the next few days and the membrane hardens. The surface retracts against the eggs and can appear somewhat wrinkled.

It is important that the whipscorpion female is not distressed or kept dry when holding the broodsac. If she is accidentally dug up while holding a broodsac it is often found the same day being ground up by her chelicerae. In the wild, this chamber is well protected from disturbances as it is normally 20″ (50 cm) below the surface, having been constructed the previous autumn when the ground was still workable due to the monsoons (Schmidt 1998). If a broodsac is formed in the open because there is no underground chamber, gently place the entire cage in a dark closet, and do not check back for two months. In captivity they can be allowed to create a burrow (beforehand), or can be given a fake burrow such as a hollow cork bark log or small coconut hut. The female spends the entire gestation with her abdomen lifted up in the air so the sac does not touch the substrate. It looks uncomfortable but

(Top and Bottom) Infertile eggs of a mature female that has been in captivity for twenty-eight months. The white eggs transformed to pink over a two-week period before the female ate the broodsac. Usually, infertile eggs are cast off and consumed after a few days.

Infertile *M. giganteus* failed eggs removed from a broodsac and compared to a 6 mm airsoft bb.

Fertile eggs can also fail and be discarded and then consumed by the female.

she takes the job seriously. Food items should never be offered because they can damage the broodsac or stress the female. Dry conditions will kill the eggs.

An unhealthy female or one kept under poor conditions can die while holding the eggs and all ova die with her. The broodsac membrane is not detached from the oviduct. It is not possible to gestate eggs without the mother. According to Frye (1992), the eggs are nourished with "false scorpion milk" as the broodsac is linked to the mother's blood cavity (haemocoel). This is certainly true at least early on since the excess fluid is visibly withdrawn back into the female's body over the days following broodsac formation. Eggs are never exposed to air and gestation is functionally similar to live bearing. Although there is no harm in attempting to save eggs from a cast off broodsac, no success is likely or has ever been reported. Even if possible, 1st instars would need to be introduced to the mother's back which probably would not work if other 1st instars were not already present.

If any eggs turn pink, they are infertile or damaged, and the entire broodsac will be cast off within a few days, less commonly a few weeks. The female will then begin to feed on the broodsac. This may be related to the

This wild-caught adult female Chinese whipscorpion lived for three years in captivity and produced an infertile broodsac each fall (no male was available). © John Skierski

quality of the eggs rather than bad mothering since clear white eggs do not seem to suffer the same fate. When eggs from a cast off broodsac are rescued and kept damp, as well as coated liberally with methylparaben to prevent mold, the eggs still do not develop.

It is very difficult to make out in the photograph, but this broodsac contains postembryos with visIble legs, no longer eggs. © Michael Fanto

Postembryos hatch into 1st instars, climb, and arrange themselves on the mother`s opisthosoma. © Michael Fanto

1st instars spend over a month hanging onto the mother and move very little. © Michael Fanto

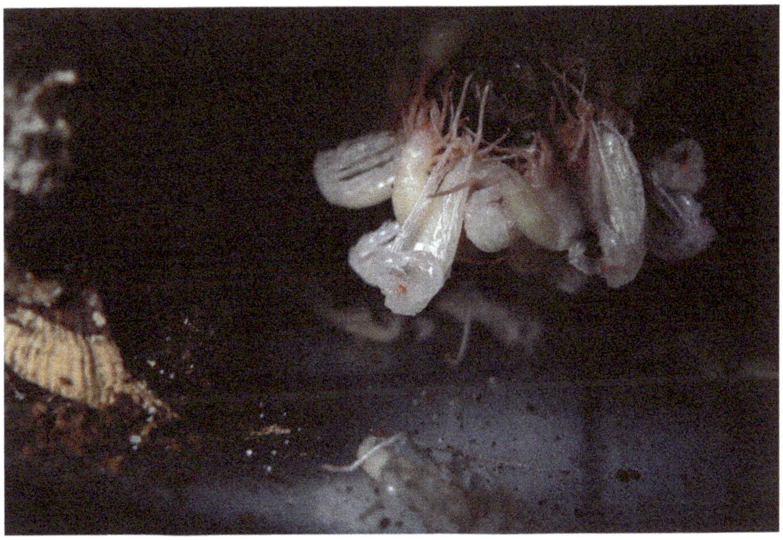

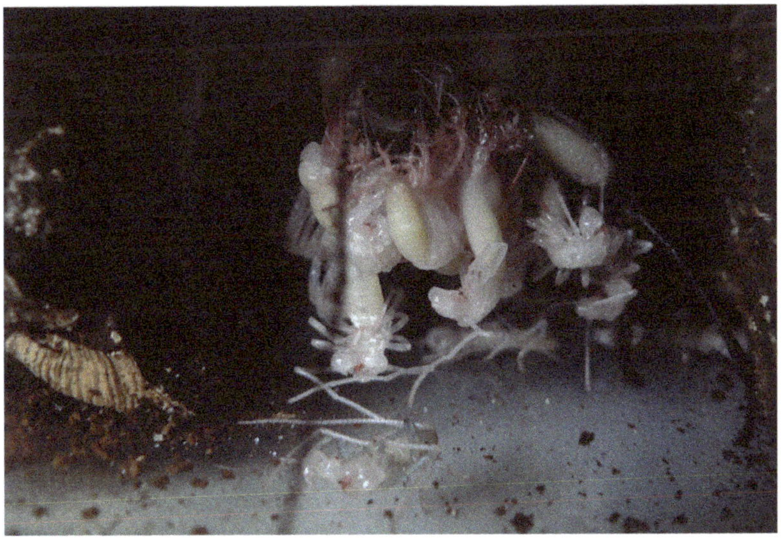

1st instars become active again as they begin to molt to 2nd instar. © Michael Fanto

Only exuviae remain on the mother. © Michael Fanto

A cork bark "cave" decorated with 2nd instar tails. © Michael Fanto

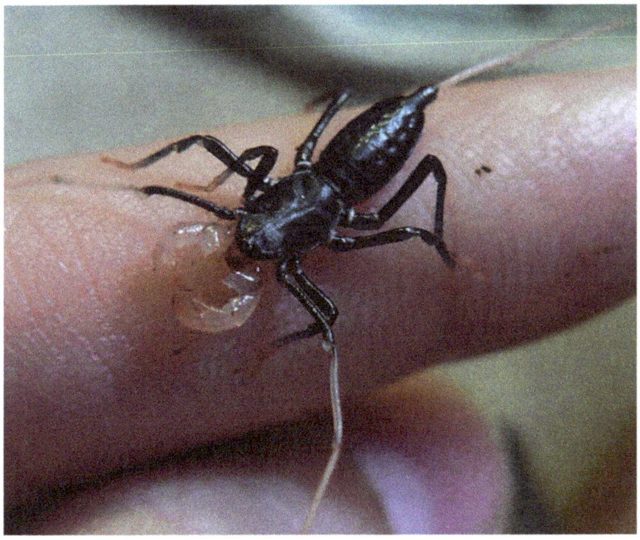

It takes days for the 2nd instar coloration to darken from pink and green to black and red. © Michael Fanto

2nd instars. © Michael Fanto

(Top and Bottom) The colorful group of 2nd instar *M. giganteus*. © Michael Fanto

TABLE 3.
MASTIGOPROCTUS GIGANTEUS
DEVELOPMENT LIFE CYCLE

STAGE	AVG. MINIMUM TIME
Internal Gestation	5-6 months*
Egg after Extrusion	3-5 weeks*
Postembryo	7 weeks
1st Instar	5 weeks
2nd Instar	1 year
3rd Instar	1 year
4th Instar	1 year
5th Instar	1 year
Adult	2-5 years

*Estimated.

The number of eggs in a *M. giganteus* broodsac is usually around thirty with maximum documented numbers of 46 for a captive-conditioned female (Fanto 2006) and 68 for wild-caught females (Schmidt 2003). Since each egg is relatively large, a broodsac takes much of the female's resources. With normal size broods, the combined weight of young (before their first meal) produced by a female in one season makes up 93% of her remaining weight (Schmidt 2003). Smaller species tend to produce a few dozen or fewer young. Sixty-eight ova is probably close to the absolute maximum egg count for any broodsac of this arachnid order. However unlikely, it is possible smaller species that have not been kept yet, produce smaller eggs in greater quantity.

Unmated females can produce broodsacs. If a wild-caught female produces a broodsac within six to eight months, the eggs will usually be fertile, since she mated the previous season. Females, however, have been kept for over two years before extruding infertile broodsacs. These were not exposed to other specimens at any time. Spiders also lay infertile eggs, but whipspiders do not form broodsacs if not exposed to a mature male (with exception of two very small parthenogenetic species).

The broodsac is held below the abdomen from less than a month for some tropical species to nearly three months for *M. giganteus*. The gestation range seems notably shorter than for amblypygids, but the overall development time can be much longer because it can take six months for broodsac formation after mating. Similarly, a third to halfway through, the postembryo stage becomes visible through the walls of the broodsac (McMonigle 2008, 2013), but in whipscorpions this stage is very easy to see because the broodsac membrane is usually clear though the shape and color transformation is usually less pronounced. The eggs will have grown noticeable but nonfunctional legs and eyes, and resemble poorly formed first instars. As with spiders, this postembryo stage is unable to move. The completed transformation is not instantaneous but becomes visible only after a set number of days, depending primarily on species, but also temperature. Development or expansion of the appendages may appear to continue for another two to three days. Little to no fluid remains within the broodsac when this stage is reached. The postembryo hatches into a first instar only after the broodsac membrane bursts, as initiated by the mother's abdominal flexing. By this time the broodsac membrane has become stretched and brittle. They hatch in waves (not all at once) which allows them to climb and encircle the mother's back in an orderly fashion. The first instar stage also does not feed, but it is more developed and capable of movement. It appears fully mobile for the first few hours when it climbs onto and positions itself on the mother's mesosoma. Once settled in place amongst its siblings, the first instar usually will not move unless prodded.

After changing to the postembryo and then hatching to first instar, the young are not yet out of the woods. The mother must also be alive for the first instars to survive. Although first instars do not feed, they die if stranded on the substrate. Even if they could survive more than a day or two without her, they would not make it through the next molt. They are simply not built to live elsewhere, though they are capable of some movement. First instars can climb and jostle to reposition themselves as other young hatch over the next few days, but are easily exhausted and incapable of locating or climbing the mother. The mother is unlikely to help young that do not make it onto her abdomen and they cannot find their own way back if they fall. Stranded nymphs should be carefully set on her back. Any that fall off after the initial settling period may be sickly or deformed and unable to hang on.

Molting to 2nd instar. © Michael Fanto

When they hatch, first instars are nearly white but develop pink highlights, especially on the legs and flagellum, which darken as they approach the molt. The first molt occurs up to a month after hatching. Although the eggs were extruded together over a few hours, it can take three days for the entire brood to molt, less for hatching. Staggered development, which facilitates initial arrangement of the first instars, becomes greater and may be a result of the limited space at this time. There is only room for a few at a time to molt successfully. Those positioned on the sides and below the opisthosoma molt from a hanging position, while those on top molt upright. Another ten days are needed for the immatures to fully darken and become ready to conquer the world. Usually, the 2nd instars are ready to venture

Thelyphonus (Tetrabalius) sp. ex. Borneo mother and young. Few other arachnids produce such relatively large offspring. © Paul Bollinger.

The palp coloration of 3rd instar is similar to that of 2nd instar. *M. giganteus*, broodsac to postembryo 5/2011, postembryo to 1st 7/2011, 1st to 2nd 5/2012, and 2nd to 3rd 6/2013.

By 4th instar the palps are nearly solid black with a red tinge. *M. giganteus*, broodsac to postembryo 3/2010, postembryo to 1st 5/2010, 1st to 2nd 4/2011, 2nd to 3rd 5/2012 and 3rd to 4th 3/2013.

Mother and baby. © Sara Pagel

off and feed shortly after the exoskeleton has fully sclerotized; a few weeks or so and the mother leaves her chamber. On the other hand, the fully colored 2nd instars of at least one *Thelyphonus* species can remain on the back of the mother after she leaves the shelter, like in most scorpions (Weygoldt and Huber 2013). In species like *M. giganteus* where the 2nd instars have bright red pedipalps, the palps do not transform to black until 3rd or 4th instar.

Mother with group of well-fed young. © Sara Pagel

The molt from 1st to 2nd instar takes only a month or so but later molts are prolonged. *Mastigoproctus giganteus* immatures only molt once per year. Although specimens are only active and feeding about five months of the year in nature, they can be kept fed and active a full twelve months of the year in captivity. Surprisingly, year-round feeding in captivity does not speed up growth and does not shorten the annual molting cycle. The speed of development for various tropical species is uncertain though the pictured male Malaysian *Typopeltis* sp. hatched in 2008 and matured in 2013. Likewise, *Thelyphonus* (*Tetrabalius*) sp. ex. Borneo required just shy of a year for the molt from 2nd to 3rd instar. *Thelyphonus caudatus* and *Typopeltis stimpsoni* molt intervals have not been documented from early development, but immatures do not molt quickly and a year interval seems to be the norm.

One reason *M. giganteus* have not been reared through multiple generations is the entire cycle is rather long. Development times (table 3) mean even if the female produces young the first year after maturity, the time between the first mating of a female and her offspring's offspring emerging is about six years. Delays can be common and will add on a few years.

There is an often recited myth that the female dies shortly after her young disperse (Stoops and Martin 1995), but this rumor is based only on short term episodes and limited data. It is true that wild-caught females which shortly produce a brood often will die, sometimes before the eggs develop. On the other hand, females that have been in captivity for many months or years before broodsac production usually live a year or two afterward. Brood production is a highly stressful and intensive period, so females exposed to shipping or collection stresses too close to this event are likely to succumb during or shortly after. Female *M. giganteus* have produced subsequent broodsacs the following spring while tropical species in less seasonal climates may not always require a full year. One female can produce two to three consecutive annual broods but more would be highly unlikely under the best of conditions.

4
HABITATS AND CAPTIVE CONCERNS

Most species are found in wet tropical forests and prefer high humidity. Some whipscorpions, specifically North American *Mastigoproctus* and African *Etienneus*, are associated with arid areas, and stay buried during the dry season, emerging with the monsoons. *Mastigoproctus* are found in seasonal dry forests and scrub rather than extreme areas of our deserts. Despite living in somewhat dry areas of the southwest, they cannot survive lack of moisture and exposure for days. With its thick armor and fat body it is surprising how sensitive to desiccation the giant vinegaroon is when compared to desert scorpions and tarantulas. However, whipscorpions are active only when surface water is available and only at night. They are excellent burrowers, which allows them to avoid excessive heat and dryness in order to exceed the boundaries of tropical forests. During the active season, existing shelters are used or adapted, such as crevices below rocks and logs or within abandoned rodent burrows. Some Indian species are found hiding during the day in termite nests (Blanford 1900).

Captive enclosures should be designed for solitary living, or short-term mating periods, unless the goal is to monitor intraspecific behavior and loss of life is not a consideration. Vinegaroons readily eat each other even if they seem to cohabitate for a time. These are not like amblypygids, where there can be a benefit to social housing (because communal caging can indirectly result in improved growth and survival). Amblypygid siblings molt next to each other safely and specific parameters can be used to induce facultative sociality from hatchling to maturity, but the same conditions would only cause cannibalism in whipscorpions. Even in the case of generally tolerant 2nd instars, there is no positive result to be gained by keeping whipscorpions together. Breene (2000) reports collecting a mature male and female *M. giganteus* from the same burrow outdoors, but within an hour of placing them in a cage together the male had eaten the female.

A critter cage or 2.5-gallon aquarium works well to house each adult or large immature; early instars can be kept in vials or cups with lids. Vinegaroons cannot climb smooth glass or plastic, so a lid is used to retain moisture more than to keep the animal from escaping. However, they can climb a coconut hut, cork bark, or pile up substrate against the side of the cage to escape. For some species, the rapid sprint of a startled animal can be enough to propel it over the walls of a shallow container it normally could not climb. Tiny immatures and tropical Asian species do well with a few pinholes for ventilation as long as large pieces of rotting prey are not present. Too much ventilation leads to excessive watering or

Thelyphonellus amazonicus habitat and specimen in situ. "I encountered single specimens at different places in French Guiana, but these photographs are from the Kaw Mountains where I have found a few specimens. I always found them under rotting logs while looking for some small theraphosids. In French Guiana, there are *no* rocks! Well, yes beside the water, but there are no rocks to lift in the forest or pretty much anywhere else. Small animals simply live in the leaf litter or under rotting logs. This species is much smaller than *M. giganteus*, I never found a specimen more than roughly two inches body length (without the flagellum). As you can see from one of the habitat shots, there was water, but I did not find specimens in proximity. I always found them in the forest, per se; it was only in the Kaw Mountains that I found a specimen where there were small water cascades, just a coincidence" (Martin Gamache pers. comm.). © Martin Gamache

desiccation, while less can cause premature death, especially for large overweight specimens. Ventilation requirements can vary greatly with seasonal home humidity levels, but at least 10% of the cover should be open for large *M. giganteus*. Tropical forest species seem to have no trouble with very little ventilation even as adults, but large immatures or adults of desert species like *M. giganteus* can suffocate over time with minimal ventilation.

Whether to choose a highly intricate terrarium or utilitarian habitat will be decided by the purpose and desire of the keeper. Huge terraria with plants, deep substrate, and various cage decorations can make a handsome display in the home, but it can be very difficult to monitor feeding, damage, and health. Also, even the largest plants can be unearthed and water features may be destroyed, filled in with dirt, or transformed into deadly pitfall traps depending on their construction. Tiny enclosures reduce variables, can prevent whipscorpions from endlessly searching for a place to hide, and are useful for individually housing large numbers of specimens. However, interaction and burrowing cannot be observed, while poisoning through dying prey or drying out if the keeper forgets to check on them for a month are more likely.

The type of substrate can be more important than the volume. Coco fiber and potting soil are substrates that work well and usually do not stick to the exoskeleton. (Wild specimens are often plastered with mud that can last

In New Mexico and some southwestern areas the *M. giganteus* habitat can look rather sparse. © John Lorbiecke

years if not removed.) Wet, pasty soils or composts can build up large clumps on the legs and lead to lethal exhaustion. Other commonly available substrates are harmless but obscure the beauty of the animals: sphagnum moss pieces are difficult to scrape off the exoskeleton, while vermiculite can completely obscure the natural coloration. Wood chips, sand, or gravel alone are poor substrates because they retain little moisture. A good portion of sand can keep substrates from becoming clumpy but also improves burrow construction. Several inches of substrate can be added to encourage burrowing, but any heavy objects should be supported or avoided entirely to prevent digging vinegaroons from being smashed. Depth of substrate can be functional to retain moisture, but mostly relates to personal taste. Hernandez (2001) stated they require more than five inches (>12.5 cm) of substrate depth, but all the photos from said article are mine and the various specimens were kept successfully in enclosures with less than 1″ (<2.5 cm) depth. Whipscorpions do not seem to know or care that they are just moving the same shallow substrate over and over again. Deeper

Alluvial forest habitat of *T. panayensis*. © Christian Schwarz

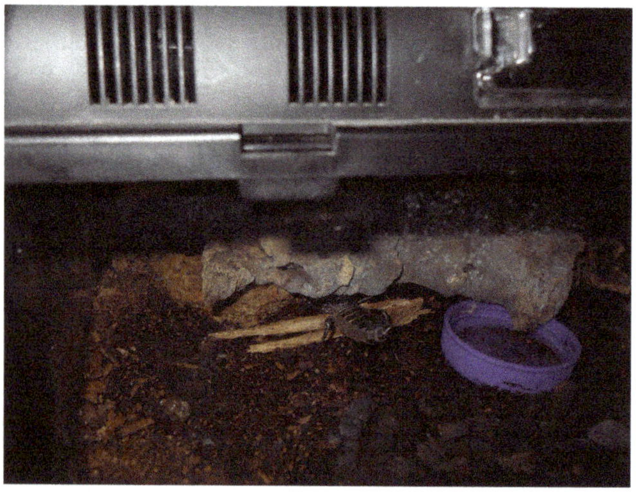

Plastic critter cages make good enclosures for *M. giganteus.* Good ventilation means a water dish should be provided. A thin layer of wood chips is adequate substrate as long as a cork hide is offered.

Typopeltis stimpsoni in 2.5 gallon glass terrarium with a glass lid. Limited ventilation is safest for most exotic species.

substrate, especially a very thick layer in vials, can help to maintain humidity, but can retain too much moisture if overwatered. Ventilation should be adjusted rather than substrate if the enclosure dries out too quickly. Tiny specimens at the most risk of drying are also the least likely to burrow deeply. Substrate that holds together when damp can collapse as it dries. This is only dangerous if the animal is about to molt or is brooding.

Use of large enclosures requires a smaller container or shelter inside, since whipscorpions like to feel objects touching them. If forced to sit out in the open in a bare tank with very shallow substrate, they can stress and die prematurely (over years not weeks). Lack of a hide may even delay molting. Common shelters include thin bark pieces from almost any tree, curved cork bark, or very small coconut huts. Over time the animals will chew small pieces off the edges of bark that is kept damp and it will eventually have to be replaced. Cork bark and coconut huts are usually immune to being

They really do drink from the water dish.

chewed up. Rocks may be sturdier and the most common hideouts in many habitats, but are overly risky and unnecessary in captivity. Large, heavy objects are dangerous and even thin slate has the ability to cause damage or death. Plastic containers or lids can be used, but should be opaque since whipscorpions prefer dark hideouts during the day.

Ready access to water is extremely important. Moisture can be provided in a shallow dish or in the form of a damp area of the substrate. If ventilation is minimal it may only be necessary to mist a section of the dirt every few weeks or months. When the enclosure is well vented, a water dish is the difference between life and death. If the water dries up rapidly, the dish may have to be refilled a few times a week. Small dishes usually get tipped over, but are preferable because oversized dishes can act as a deadly pitfall trap. If the substrate is shallow the water dish is less likely to be filled with dirt or flipped over.

Full-grown Costa Rican "jungle micropods" may be useful. Other commonly kept clean-up isopod species are usually big enough to be eaten.

Room temperature around 75° F (24° C) seems to work well but 70°-80° F (21°-27° C) is a reasonable range. *Mastigoproctus giganteus* can survive below 40° F (4° C) for hours,

Springtails are tiny, harmless, and useful for clean-up, but a little overbearing.

possibly days. Accidentally frozen shipments remain motionless for up to 48 hours and specimens almost always recover. However, gravid frozen specimens do not recover, or extrude a dead egg mass within a few weeks. Tropical species would probably die rapidly below 55° F (13° C), but I can find no data on frozen shipments and I am not sure anyone would consider purposefully testing the lower limits of creatures that are so hard to find. Whipscorpions are fully nocturnal and hide in crevices during the heat of the day. They may survive incredibly high temperatures, but there is no known reason to supply scorching heat or intense sunlight to a captive habitat.

The most common problem in captivity other than desiccation is associated with the hypopus stage of grain mites caused by overfeeding and moisture. These tiny mites are found nearly everywhere, and are prevalent in cricket and fruit fly cultures. Following a population explosion, the mites stick to the body and can restrict breathing channels. Substrate should be discarded and the cage and decorations thoroughly washed. Small numbers of hypopus stage grain mites are not harmful and it can be safer to let them dwindle over time

Cohabitation Dramatization: In truth the smaller specimen is a 2.5 year-old, 4th instar *M. giganteus* that died mysteriously and was immediately taken by this adult *T. stimpsoni*. Had they been the same species and both alive the results would have been similar.

Predatory mites like this *Hypoaspis* sp. are extremely small, the size of a fruit fly eyeball, but still are larger than most grain mites. © Henry Kohler

with improved husbandry than assault them on the whipscorpion. Removal of leftover and dead prey along with reduced feeding and a "clean-up crew" of large springtails or small isopods will prevent an outbreak. A clean-up crew is usually not enough to compensate for excessive overfeeding since it removes limited amounts of waste. Predatory mites are useful to remove mites attached to the animal's exoskeleton, but it can take a month or two and culture starts are inconsistent and expensive. Predatory mites like clean-up crews will not compensate for excessive rotting prey. Even if they could, byproducts of bacterial decomposition would suffocate or poison the animals.

A little bit of mold can be a sign of husbandry problems but is generally harmless on its own. Mold growth or mushrooms that arise from damp substrate or wood are harmless. On the other hand, molds that grow on dead prey can cause illness or death relative to the amount of mold, ventilation, and surface area. This should be prevented by removal of excess prey and clean-up crews. Rarely, tiny flecks of mold show up on the surface of the whipscorpion's exoskeleton, primarily on the legs, but it does not seem to have an effect, like algae growing on the back of a turtle. It can be gently scrubbed off, but excessive "attention" might stress or kill the animal.

A main reason for keeping vinegaroons is they are great animals for handling and display since they appear quite scary to the average person and yet are harmless. However, handling is not always harmless to the animals. Big, fat females easily crack if dropped on a hard surface, so they should never be held high above a surface, especially not cement or tile. Careful and immediate application of super glue may prevent some blood loss, but will not help massive damage and can do more harm than good if it blocks the entrance to a book lung. Phorid flies and grain mites can attack a large open wound, even if the animal does not die right away and the wound has dried. Thin males and immatures that are not close to a molt can survive a pretty bad fall, but may be found dead the next day.

Coconut huts make good shelters. © Sara Pagel

This chapter covers a number of possible problems to prevent unnecessary deaths of the animals but it is not meant to discourage. Those who are already familiar with caring for any small animal will likely find vinegaroons are the easiest things they have ever kept. Vinegaroons are difficult to breed but require little effort to keep alive.

5
PREY AND FEEDING

Among arthropods, the whipscorpion is an apex predator that eats spiders, scorpions, windscorpions, assassin bugs, and praying mantids, but also any arthropod that moves and gets too close to the palps. None of these arthropods pose a threat to mature *M. giganteus* and even the immatures are protected from most attacks. *Mastigoproctus giganteus* from the American Southwest and Florida have no known natural predators. Studies of population dynamics show that adults are not killed by other animals and the main limiting factor is prey availability (Schmidt and Schmidt 1998). The spray is so effective that even the most voracious predators (i.e. solifugids, lizards, and birds) are repelled. More powerful secretions and faster movement of some exotics suggest vertebrate predators pose a threat in their habitats.

As a whole, representatives of this order will feed on a great variety of invertebrates and are not perturbed by many, if any, defensive poisons or venoms. Darkling beetles, millipedes, orthopteroid insects (such as grasshoppers and crickets), and scorpions are the most commonly seen prey of *M. giganteus* in its southwestern range. Flies and moths are also abundant and harmless, but easily escape capture through flight and likely compose little to no part of the natural diet. This plays out somewhat in captivity as harmless flies are often ignored while the heavily protected but slower scorpions, thick-shelled wingless cockroaches, and armored beetles are attacked with zeal. Punzo (2000) studied wild *M. giganteus* in Big Bend National Park in Texas and found the most common prey in this habitat included 17% spiders, 23% orthopterans, 14% beetles, and 14% true bugs. Most true bugs are not predatory and therefore not venomous, but the herbivores host an array of powerful chemical defenses. Vinegaroons easily overpower venomous arachnids in captivity though their importance in the diet is uncertain. Evans (2007) claims the giant whipscorpion feeds on arthropods ". . . especially scorpions and windscorpions." This may only be an inference from the most conspicuous, nocturnal prey common in the same areas since no reference or data is provided. The data may have come from Justin Schmidt's (2003, 2009) observations of the feeding habits of *Mastigoproctus giganteus* in Cochise County, Arizona. He discovered they consume a large number of scorpions and windscorpions, as well as spiders, beetles, caterpillars, crickets, and various insects. Beetle prey included the heavily protected blister beetles and *Calosoma* ground beetles. (Humans tend to regret picking up either of these, let alone placing one in the mouth.)

The only feeding problem is whipscorpions eat almost anything that moves and some things that do not. Feeding on immobile prey is not an artifact of captivity—it has been observed in nature. *Mastigoproctus pelegrini* was observed feeding on the eggs of a snail, *Zachrysia guanensis*, and eggs of a terrestrial frog, *Eleutherodactylus* sp, while an unidentified *Mastigoproctus* was photographed eating a lizard egg, *Anolis* sp., in Cuba (Armas et al. 2009). The propensity to eat almost anything would seem to make husbandry easy, except they can eat prey or other items which may be dangerous or nutrient deficient. Cultured prey is recommended since it is usually possible to avoid introduction of disease, parasites, or pesticide residuals. Some chemically protected insects may be harmful if used as a regular part of the diet. Considering the long periods between molts it can take years to fully discover problems caused by specific diets. Sato (1941) fed his *Typopeltis stimpsoni* mostly on three different fly species (and at least one *Lycosa* wolf spider as well as feeding his *Typopeltis* to each other) in the lab for short-term experiments, but whipscorpions could be fed sugar jellies or nothing at all for months and seem fine. A wholly improper diet could lead to problematic molts or explain molting delays, gestational failure, and unexplained deaths. Just being fat and well fed does not necessarily mean healthy. They attack a variety of prey and should be provided varied quarry for best results.

Whipscorpions have a powerful appetite and can easily overpower prey as large as themselves. They do not require huge prey items but seem to prefer a challenge and will ignore very small creatures. Second instars can be started on standard isopods and mealworms their own length, but can starve if offered only tiny prey such as springtails or small fruit flies. Starving animals may attack prey much smaller or much larger than normal. Hunger is usually related to how recently they have fed, as it is barely perturbed by molting or broodsac formation. Hunger also relates to gender, as adult females will consume a much larger portion of a massive prey item than a mature male will.

Although it may be possible to feed them entirely on dead insects or pieces, live prey is suggested in most cases. Freshly dead prey may be used and the liquefied contents are readily consumed by hungry animals. Small immatures will feed on freshly squished adult crickets if small food is not available. Canned, dried crickets or dried krill seem to be wholly ignored (starvation testing was not performed). Dead prey is easily covered or buried so that feeding is difficult to monitor and lethal buildup of rotting material becomes likely.

Unlike many pets, vinegaroons can go a few weeks between feedings with no ill effects and may even be healthier when not fed too often. In nature, *M. giganteus* probably only feed for about five months of the year. Feeding more than twice a month is usually not necessary as long as the prey is as big as the uropygid carapace. Wild-caught animals that arrive emaciated should be fed a few times a week until they look healthy. Females commonly have huge round abdomens but males can be made to look almost the same in captivity when overfed. Allow the abdomen to thin out slightly between feedings. They almost always overeat in captivity if allowed. Only females and immatures should look especially fat. Females with broodsacs will usually still feed but this can lead to the broodsac being cast off and eaten. Heavily fed whipscorpions are more likely to break open from a short fall. Obese animals can die from reduced ventilation and yet are not proportionately protected from desiccation.

Burrowing prey is taken as readily as surface prey. Burrowing mealworms or cockroaches like *Blaberus* and *Eublaberus* work

This Japanese whipscorpion is eating a Japanese beetle, *Popillia japonica*.

Feeding on a mealworm beetle, *Tenebrio molitor*. © Michael Fanto

Medium-sized *Blaberus* nymph being consumed.

well as food while glass-climbing species such as lobster roaches (*Nauphoeta cinerea*) can avoid capture indefinitely unless the cage is very tiny and shallow. Earthworms are often eaten in a small enclosure, but can burrow, climb, and hide in narrow cracks to escape. Fast moving crickets and high profile roaches like *Blatta* also work, but escape predation easily if they can climb onto a cage decoration. Whipscorpions are not shy feeders and will usually be chewing apart the prey within a minute of it being tossed in. Unlike whip-spiders, they almost never toss the prey out due to minor disturbances, and would come back and finish it later anyway.

A range of beetles are available as prey cultures. The most widespread is of course the yellow mealworm *Tenebrio molitor*. Whipscorpions cannot always hunt down the burrowing mealworms but the adult beetles make good prey for medium to large animals. *Eleodes subnitens* is an easily cultured giant darkling from the natural range of *M. giganteus,* but it is more difficult and slower growing than *T. molitor*. Second instar whipscorpions readily take adult pea weevils *Callosobruchus maculatus,* and will be happy to take down larvae of larder beetles *Dermestes lardarius* and hide and tallow beetles, *Dermestes talpinus*. However, they are not likely to tackle these large, hard-shelled adult dermestids until 3rd or 4th instar.

This male vinegaroon attacked a *Chicobolus* millipede from under the head and was feeding on it despite defensive secretions. It ran off with the head and continued to feed when the lid was opened for a photo.

Isopods are not only good as clean-up crews, but have proven highly valuable prey items in captivity. I have had a few sickly imports that refused other prey until they had time to eat some isopods and recover—following a few weeks or months they accepted other prey. They would not accept crickets or cockroaches and would likely have starved since they arrived emaciated.

Giant millipedes are known to be part of the wild diet (at least for *Mastigoproctus giganteus* found in the Lake Wales Ridge area of Florida). Some were seen eating *Floridobolus penneri* in the wild, though most cases were from the millipedes being trapped together with a whipscorpion in a pitfall trap (Carrel and Britt 2009). *Mastigoproctus giganteus* specimens have eaten specimens of the very large spirostreptid millipede, *Orthoporus ornatus,* which occurs in their western range. This has only been seen in captivity since reports of natural predation are scarce. *Mastigoproctus* do seem unaffected by the strong quinone defense of spirobolid and spirostreptid millipedes, but darkling beetles, a common prey, also employ quinones in defense. Large millipedes are rarely used because culturing as a food item is hardly realistic. Wild specimens not only tend to be expensive, but may have been exposed to disease or pesticide. Polydesmid flat millipedes can be easily collected in some areas, but their secretions contain cyanide and may not be so benign over time.

Feeding on a beetle grub. (Someone is sharing with Mom.) © Michael Fanto

Typopeltis feeding on a large mealworm of the darkling beetle *Eleodes subnitens*.

Phasmid being consumed.

Earthworms and slugs are listed as primary foods by Stoops and Martin (1995). Vinegaroons have eaten earthworms in captivity but the worms are not native and cannot be found in the southwest. Still they can be picked up from the local bait shop if no other foods can be found. Worms coat themselves in dirt, climb glass, and escape feeding even in the smallest enclosure. They are also prone to dying at warm temperatures. Slugs leave slime trails and would have to be wild-collected.

Wolf spider (*Lycosa* sp.) being consumed.

Two of the most common natural prey, other arachnids and true bugs, are seldom employed in captivity. Spiders and scorpions are not likely to be useful feeders in captive situations unless locally abundant and the area of collection is reasonably certain to be pesticide free. Culture efforts and required time are disproportionate to their use. Another problem is they may envenomate the keeper during capture or feeding. True bugs likewise make poor feeders. Assassin bugs (specifically the easily cultured African *Platymeris* spp.) might present a hazard if they bite the whipscorpion. Only the large milkweed bug (a rather small bug overall) is readily cultured but it can be harmful to mantids, giant water bugs, and assassin bugs that feed on it.

Considering adult size, strength, voracity, and habit of entering foreign burrows, vertebrate prey in the form of small mice and reptiles are likely eaten by adult *M. giganteus* from time to time in nature. One study found that 25% of the *M. giganteus* population took shelter in abandoned rodent burrows during the day (Punzo 2006). Vinegaroons do scavenge, so dead mice likely make up a small part of the natural diet, even if young mice are never separated from the mother and so are protected from attack. Various authors reference the use of pinkie mice as food (Frye 1992, Bruins 1999, Fanto 2006). Some successful breeders feel it is helpful to offer a pinkie to large females after mating but there is no data to support or contradict this experience. In this case live prey is not suggested because live mice have sharp teeth and they scream.

An adult male *M. giganteus* feeding on a pre-killed fuzzy mouse. In the case of oversized meals, much is discarded and leftovers should be removed as soon as possible. © Michael Fanto

Gut loading and dusting are methods used to improve the quality of the prey item. Gut loading involves feeding the prey item high protein dog food, powdered milk or other highly nutritious item prior to feeding the arachnid, which then eats the prey and at least partly digests its gut content. Various powders containing vitamins and minerals are available and designed for dusting prey of insectivorous lizards. There have been no controlled experiments, but better nutrition may improve survival or growth. Dusting can measurably affect arachnid and insect size (McMonigle 2006) since it can make up for deficiency in the chosen prey.

The hobbyist who keeps more than a few dozen specimens or wishes to captive breed will almost certainly want to culture at least a portion of the necessary prey items. Selecting a specific set of cultured prey items is especially important when rearing immature whipscorpions, since it reduces variables and makes it possible to evaluate the diet. Rearing prey requires an investment in time and space but can save a lot of time and money in the long run. It can be difficult to get small food items of the appropriate size from local shops to feed immatures. Culturing your own prey assures in some cases the animals are not wild caught (i.e. pillbugs), malnourished, or carrying diseases. The following prey items are commonly used, readily reared, and proven foods.

EASILY CULTURED PREY

Cockroaches—There are numerous available feeder species. Those that cannot climb glass, such as the discoid roach *Blaberus discoidalis* and *Blaptica dubia,* are useful for medium to large whipscorpions. Discoids grow large, so a half grown nymph is a huge meal for the largest vinegaroon. They are very useful for vinegaroons because they do not climb glass and are born small enough to be fed to 3rd instar vinegaroons. Lobster roaches, *Nauphoeta cinerea,* are a glass climber but they tend to hang out in the substrate and are very easy to culture. Also, newborns are small enough to feed any 2nd instar vinegaroon. Other live-bearing roaches can work well, but many are slow growing or less easily cultured at room temperature. Colorful, high profile, egg layers like *Dorylaea orini* and ornate velvets, *Deropeltis paulinoi,* can also be used but they climb glass and easily stay out of reach indefinitely. *Deropeltis* and *Diploptera* species emit a quinone chemical defense like darkling beetles and giant round millipedes. Pest species might be useful food, but tropical pet species are normally the only ones raised as feeders because they cannot survive long in a human structure.

Pros: Roaches do not attack molting whipscorpions (with the possible exception of *Eublaberus posticus* in quantity). Many cockroach species (primarily the live bearing Blaberidae) are very nutritious. They do not jump and do not chirp. They can survive weeks or months in the cage until the vinegaroon is hungry. A decent colony produces a far greater volume than crickets in the same amount of space. Most people can culture these with very little time or effort.

Cons: Cockroaches are primarily sold as culture starts or cheaply in large batches, so it is necessary to maintain a colony, which may only be worthwhile for enthusiasts owning quite a few animals. Colonies can take some time to get going.

Rearing Instructions: A ten-gallon tank or five gallon bucket with screen lid makes a good cage. Discoids (as well as other *Blaberus, Eublaberus, Archimandritra, Blatta* and a few other genera) cannot climb smooth surfaces. Most roaches climb glass or plastic so the top three inches of the sides should be coated with a thin layer of petroleum jelly to prevent escape. An inch of vermiculite or potting soil makes a good substrate. Add a sponge-filled water dish on one side and dry dog food,

Late instar pallid cockroach nymph being processed.

Earthworm being eaten.

A large isopod (*Porcellio dilatatus*) being eaten by a small whipscorpion.

2nd instar *M. giganteus* playing tug of war over crickets. © Sara Pagel

chicken mash or other pet food on the other side. Keep the food side bone-dry to prevent mold and mites. Feed only as much as will be completely consumed in less than a week. In a cool, damp basement a radiant ceramic lizard heater or 60-watt light bulb can be placed over the dry side of the 10-gallon to keep the dog food from molding and provide warmth for egg development. Occasional slices of fruit, especially apple, pear and peach often improves reproduction for live bearers.

Crickets—Most enthusiasts use this feeder because it is convenient to pick up at the local pet store. Without the regular availability of crickets the arachnid hobby would never have gotten off the ground.

Pros: They jump around to attract attention and do not burrow to escape. Priced and commonly available for direct feeding.

Cons: Uneaten crickets can chew holes in molting whipscorpions; otherwise they die and rot in the cage. Crickets should be supplemented with other prey because they are not especially nutritive. Escaped males chirp loudly and can keep the keeper or spouse up at night. Tiny crickets smash easily when handled. Adults can be even more difficult to catch and maneuver because they jump well. Sometimes they are unavailable in hot summer months because they cannot survive extremely high temperatures.

Rearing Instructions: Usually at least two ten-gallon aquariums or tubs are needed to rear crickets. A few large pieces of egg carton are added for molting and hiding. Adults are placed in a cage or tray with 1″-2″ of potting soil. An area or side of the soil should be kept constantly moist to keep the eggs from drying up. Temperatures in the high seventies to high eighties (25°-30° C) are necessary for the eggs to hatch. At normal home temperatures a 60-watt incandescent bulb or radiant ceramic heater can be placed directly over one side to provide the necessary heat. After two to four weeks, tons of tiny hatchlings emerge. Cricket chow or cat/dog food is an adequate staple diet. Fruits, vegetables, and vitamins can be included in hopes of making them more nutritious.

Mealworms—*Tenebrio molitor* are commonly available at pet shops and for bait. (The huge, dark-banded *Zophobas morio* 'Mundoworms' should be avoided if possible, because uneaten they pose a threat when the whipscorpion molts.)

Pros: They seldom die in the cage if uneaten and do not attack molting animals. Mealworms have a very long shelf life; they can be stored in the refrigerator for months. Cultures are low maintenance.

Cons: Cultures not kept extremely dry can be overrun with grain mites. It can take a very long time to get a culture started.

Rearing Instructions: Fill a small animal cage halfway with oatmeal and throw the mealworms and beetles on top. Unless an outdated tub of mealworms is purchased, it will take a few months for pupae, then adults to show up. A small chunk of potato can be added every so often to provide moisture but may provide enough moisture for grain mite problems. They cannot climb smooth surfaces but a lid is necessary since the beetles can fly. (Flight is rarely witnessed but *T. molitor* fly well.)

Isopods—*Pros:* Isopods can be collected in most areas and from under logs even when there is a foot of snow on top. They are easy to culture and have no unique requirements. "Micropod" species like *Trichorhina tomentosa* and the smaller Costa Rican micropod start out smaller than a fruit fly eyeball and could be important for 2nd instars of very tiny species. This is one of the few terrestrial arthropods with a calcified exoskeleton, but the shell is not eaten. They may be eaten by picky feeders or sickly animals.

Cons: Even large *Porcellio* are small compared to an adult cricket. Reproduction is rather slow since most large species take four to twelve months to mature.

Rearing Instructions: An inch or so of damp potting soil is placed in a plastic shoebox or bucket with a lid. A few holes can be added to the lid for ventilation together with some brown, decayed hardwood leaves and pieces of wood or bark placed on the substrate. Dog food, fish flakes, or fruit can be added to supplement the diet, but without the leaves and wood females seldom produce young.

Pinkie Mice—May improve egg production or growth for *M. giganteus* and other large vinegaroons.

Pros: Large, readily accessible prey item. Frozen (thawed) are acceptable and preferable since they do not cry or scream.

Cons: These need to be fed out right away and would take excessive space and time to rear. Cost can be many times that of a similarly sized roach.

6
DISCUSSION OF SELECT SPECIES

MEASUREMENTS USED

Standard measurement of whipscorpions and scorpions is from the front of the head (leading edge of the prosoma) to the end of the abdomen (terminal pygidial segment containing the anus) and applies to invertebrate measurement in general. (Legspan, whipspan, and antenna length should not be expressed as "body length.") Vernier calipers should be used if expressed to the millimeter or fractional inch. Body length measurements listed here use standard methods.

Measurement from the front of the prosoma to the last abdominal segment containing the anus. This female *Typopeltis tarnanii* was imported in 2010 and died fall of 2013, a few weeks after cannibalizing her broodsac.

When discussing an animal it always seems important to understand and convey the size. This can be a big problem when the way in which an animal is measured is different according to author or person. If appendages are included, or the measurement is legspan, the results should be labeled accordingly. Authors such as Bruins (1999) list the maximum size at 130 mm (5.2″) and this certainly included the flagellum, or palps, but there was no explanation or reference, and was possibly taken from Milne and Milne (1980). Most people simply do not use measuring devices and their estimate is never smaller and seldom accurate. Another major hurdle for communication is people exaggerate even if they do not want to. Even honest fishermen hold their fish out in front of their bodies for the camera so *their* fish does not look smaller than everyone else's. It makes it difficult for enthusiasts, nature lovers, and even vendors to convey accurate size to another person.

SPECIES

Mastigoproctus giganteus (Lucas, 1835)
Giant Vinegaroon

A movie monster-maker could not create a more unreal looking creature. The huge body and slow gait make them perfect for handling and displays. Large animals seem like they should have little trouble giving a moderate pinch with the pedipalps, but are very docile. (A pinch would only result from purposeful

Mastigoproctus giganteus. © Michael Fanto

goading or cruel handling.) Despite incredible visual appeal, large size, and harmless demeanor, they are one of the least familiar pet arachnids. Though giant vinegaroon is the most common name, in some areas it is called *grampus* (Milne and Milne 1980), which is a name also used for killer whales, hellbender salamanders, and hellgrammites (dobsonfly larvae).

The handsome grin of *Mastigoproctus giganteus.*

This, the largest of the whipscorpion species, occurs in Florida, and from central Texas through southern New Mexico into the southeastern oak zone of Arizona (Werner and Olson 1994) and adjacent northern Mexico. There are two records from Utah and one from Oklahoma. The Florida populations were separated by some distance and described as a subspecies, *M. g. floridanus*, but morphology does not agree with this. The only possible difference is a slight variation in mating timing (Punzo and Reeves 2001), but slight behavioral difference is not enough to support a subspecies designation. There are two other subspecies, *M. g. scabrosus* and *M. g. mexicanus*, found in Mexico.

This is often considered a rare and uncommon desert dweller because of its subterranean and nocturnal behavior. However, through mark-recapture methods in Arizona, Justin Schmidt (2000, 2009) has shown the results translate into 68,300 vinegaroons per square kilometer in ideal habitat (for an estimated combined weight of 212 kilograms per square kilometer). *Mastigoproctus giganteus* is the most important arthropod predator in much of its southwestern range.

Adults are outfitted with black armor and are usually all colored the same. Wild specimens often look dark brown since they are pasted with bits of dirt. There is usually no real variation in the black color of the hardened body segments. However, there is a rare red form that was found in El Paso, Texas, in 2012. Construction on a property unearthed just over

July 2013. This red form *M. giganteus* was collected in Aug 2012 and molted May 2013. It was one of two red specimens, out of 400, unearthed (and rescued) during construction on property in El Paso, Texas. © Traci Roach

This 2nd instar captive hatched specimen is nearly a year old. The bright red pedipalps of young *M. giganteus* can be kept through fourth instar.

Mastigoproctus baracoensis, Cuba. © Michael Seiter

400 *M. giganteus* specimens, primarily immatures. Fortunately the owners called someone to remove the specimens rather than kill them. Human concern for arthropods during construction projects is exceptionally rare, but what they unearthed may be just as uncommon. Of the 400, all were normal colored except for the two red form specimens pictured here (Traci Roach pers. comm.).

After thunderstorms, during warm spring and cool summer nights in the southwest, enthusiasts can find specimens by driving slowly late at night and scanning the roadway for roving *M. giganteus*. The number of specimens smashed on the roads in one year is probably greater than the total number ever kept in captivity.

Giant vinegaroons rarely flee and even the 2nd instars rear up and hold their ground when confronted. The incredibly slow gait makes them a great addition to a live animal show. They will not bite or sting and are unlikely to be dropped due to their own movements.

The giant vinegaroon is a very hardy captive that loves to dig and move earth around. Cages do not require substrate depth (1/4″ of potting soil is fine) but the animals do seem a little happier if given the opportunity to create burrows. Specimens can move a lot of dirt so it is important to avoid large, unsupported rocks or other heavy decorations in the cage. Heavy items can turn a harmless cave-in deadly. In the wild they are commonly found in pre-existing holes (Werner and Olson 1994), with about a quarter found in abandoned rodent burrows (Punzo 2006). This behavior takes place during active periods. Overwintering and brood chambers must be constructed. Good ventilation and limited humidity are preferred by large specimens, so substrate should not be kept too damp.

Weygoldt (1968) mentioned the males were rather rare (he had collected one male and six females in Florida). I have seen dozens of groups from Texas, usually from six to twelve specimens, and some groups have been 50% male. Females are more common and if there are only two or three adults they are often all female. However, it is extremely difficult to tell a subadult male from a small adult or subadult female. In the wild, mating probably takes place only in late summer to early fall, but they can be successfully mated year round in captivity. Females extrude the broodsac in March or April and a little over three months later the young emerge with their mother and venture off; some may remain together in small groups for the first growing season. *Mastigoproctus giganteus* females extrude the broodsac as late as June in captivity. The immatures undergo five postembryonal instars and no matter how well fed, they only go through one molt a year. Wild-caught adults have lived as long as five years in captivity.

Natural Report: "The *M. giganteus* here in Florida seem to exist in two separate habitats. They can be found on barrier islands and in the highest elevation xeric scrub habitat (which were islands at one point in time). They appear to be absent from any other habitat in between, likely due to periodic flooding in lower elevations and vast expanses of wetlands that prevent distribution.

"The population density of *Mastigoproctus* in Florida is nowhere near some areas in West Texas and New Mexico. The most we (Paul Bollinger and I) ever encountered was twenty-five in a day with the average usually around two specimens. Another collector in Florida, who is really good at finding them, reports similar numbers, though his average is better than ours. We have never seen a vinegaroon crossing a road at night in Florida, and that covers at least fifteen years of road cruising for snakes, yet they are a fairly common sight out west crossing roads after heavy rains. There are two people we know in the desert southwest

Mastigoproctus pelegrini, Cuba.
© Michael Seiter

Mastigoprocus pelegrini from Cuba; these are too glossy for flash photography.
© Martin Gamache

Thelyphonoides panayensis, Research Station Sibaliw, Panay, Philippines. © Michael Seiter

Thelyphonoides panayensis, juvenile in captivity. © Christian Schwarz

Thelyphonoides panayensis close-up. © Christian Schwarz

Thelyphonida sp. ex. Cat Ba Island in northern Vietnam. © Michael Seiter

who can easily and consistently locate a hundred or more in a single night after heavy rainfall" (Sam Floyd pers. comm.).

Size: Just under two inches (<50 mm) is common for adult specimens, but they can vary widely. The range for females is 37-65 mm and 45-76 mm for males.

Brood Size Actual Counts: 22-68.

Thelyphonoides panayensis Krehenwinkel et al., 2009

As the name suggests, *T. payayensis* is found on the Philippine Island of Panay (-ensis means "originating from"). This genus with its only species was described in 2009 according to differences in the genitalia. It is larger than *Thelyphonus* spp. Adults specimens can reach 47 mm versus 35 mm for the largest *Thelyphonus* spp. (Krehenwinkel et al. 2009). Though natural enemies are unknown, remnants are commonly found in the stomach contents of the invasive cane toad *Rhinella (Bufo) marina*. Only two or three immature specimens have made their way to Europe so far (C. Schwarz pers. comm.).

Thelyphonus caudatus (Linnaeus, 1758)
Lesser Malaysian Vinegaroon

Thelyphonus is by far the largest whipscorpion genus with fifty or so representatives. This species, *T. caudatus*, is nearly black and a mature adult is only around 30 mm. Specimens do well with high humidity and can dry out easily. Adults proved to be rather hardy for various keepers. I was only able to acquire one adult male that lived for just over three years. Similar-sized adult specimens are quite tolerant of being kept in groups when adequate food and shelter is provided. It is a capable burrower but prefers to dig very shallow scrapes under objects rather than dig tunnels. Specimens are more active than *Mastigoproctus* and tend to

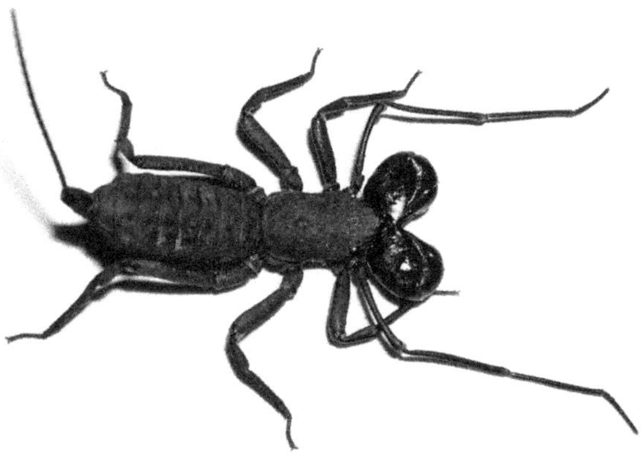

Thelyphonus caudatus adult male.

run away when disturbed. This small species was imported in numbers in 2007. This is one of the most widespread whipscorpion species as it is found from southern India, east to the Philippines, and across the whole Indo- and Austro-Malayan area as far as the Solomon Islands (Blanford 1900).

Males have a notably large genital operculum and a shorter flagellum. While the male's palps are relatively larger, it is not especially notable. Females have a tiny extra spine on the basitarsus of the whip, but this feature is very difficult to see unaided. Captive mating has been seen (as pictured here) but no broodsacs were reported.

A much larger species (about 2/3rd the size of *M. giganteus*) with pale banding between the plates on the opisthosoma was imported from Malaysia in some quantity in 2000. These Malaysian species are the only vinegaroons, other than *Mastigoproctus giganteus*, I have seen on U.S. vendor lists between 1993 and 2013, but they came and went quickly. There are a number of different species from at least two genera, so pinning down the species might even be difficult if specimens were still available. Vinegaroons are common creatures in Southeast Asia, specifically Malaysia, and visitors often run into them on holiday (Cole 1998).

A

B

C

D. *Thelyphonus* cf *caudatus* molt sequence. © Martin Gamache

Thelyphonus (*Minbosius*) *manilanus* Koch, 1843
Limited numbers of these interesting vinegaroons from the Philippine Island of Luzon made their way into Europe, and to a limited extent into North America, between 2007 and 2008. These can grow to the size of a small *M. giganteus*, and can be kept in much the same way. They look a little different because the tergites on the abdomen are divided and many body surfaces are smoother. The male's operculum is larger and pleural membrane narrower than the female's, though the difference in the palps is not as great. One interesting thing about this species is the spray does not smell like vinegar. It remains to be seen if acetic acid is a component.

Typopeltis stimpsoni (Wood, 1862)
Amami Whipscorpion or "Ninjaroon"
This handsome vinegaroon is a good-sized animal but probably never exceeds two inches (<50 mm) in body length. The defensive fluid is produced in quantity and the spray is visible even with small specimens (a slightly scary proposition without safety glasses). The animals run when confronted by other specimens placed in the same cage and can take off running when startled by the keeper. If held up on a large piece of styrofoam or cork bark, it runs around the surface faster than most amblypygids. Although vastly more mobile and aggressive than other vinegaroons when confronted with a threat, it is more timid in

Thelyphonus manilanus male.
© Michael Seiter

Males and females of some species can have similar looking palps. *Thelyphonus manilanus*, dorsal view of pair, female at left. © Michael Seiter

Thelyphonus manilanus, ventral view, male at right. Note the larger genital operculum. © Michael Seiter

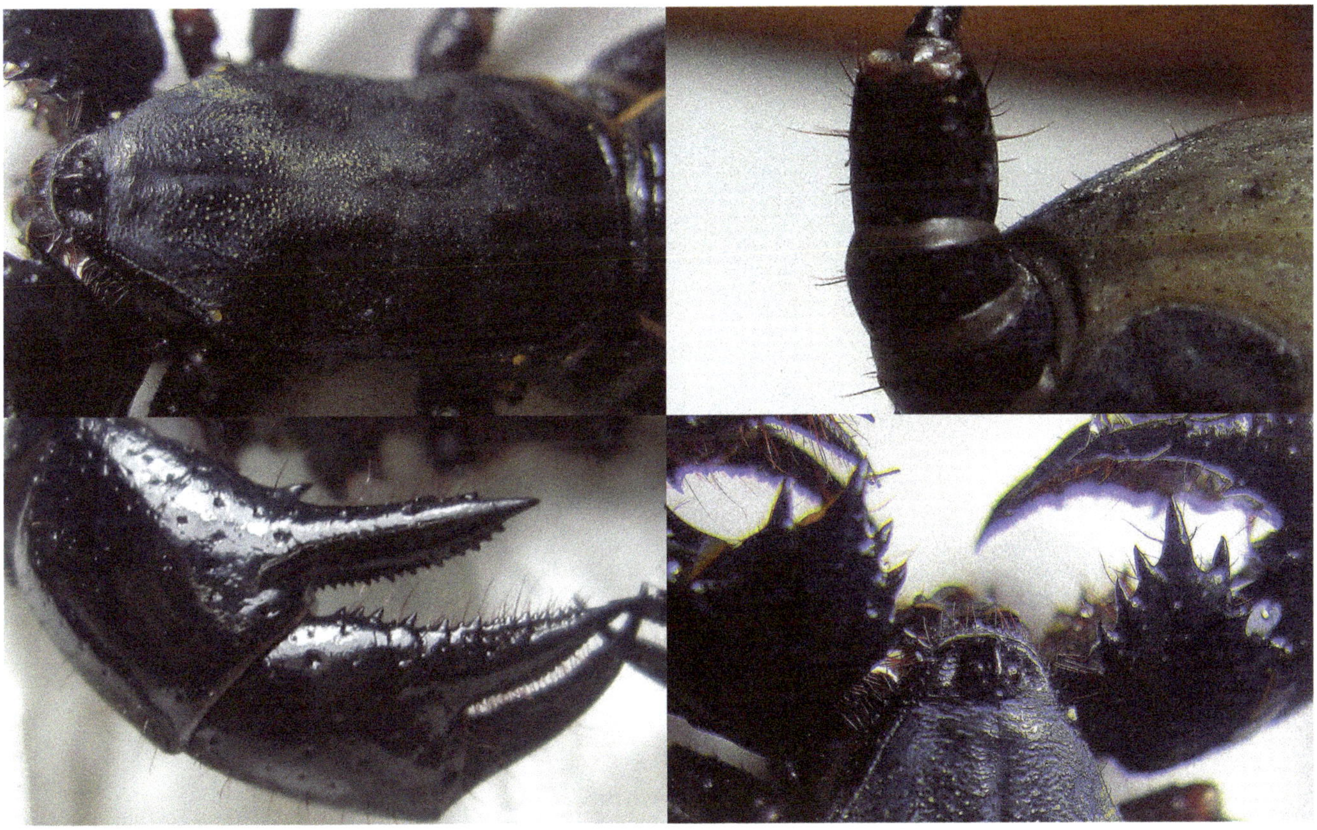

Thelyphonus sp. Kuching, Sarawak, Borneo © Michael Seiter

Thelyphonus sepiaris from Tamil Nadu south, India. © Michael Seiter

Thelyphonus sp., Malaysia. © Michael Seiter

Thelyphonus sp., Lombok, Indonesia. © Michael Seiter

Female with broodsac. *Thelyphonus* sp. ex. Northern Laos. © Michael Seiter

Thelyphonus sp. from northern Laos. © Michael Seiter

Thelyphonus sp. from Malaysia. © Michael Seiter

Typopeltis stimpsoni male.

relation to prey capture. Despite speed, specimens seldom immediately grab prey tossed in the cage like *M. giganteus*. This species is more cautious and often waits at minimum a few minutes. Wild specimens can be found in holes dug in the red clay of hillsides or beneath stones; normally only one is found in a hole (Sato 1941).

In this and a few other genera the males have an especially long palpal apophysis with a bent, flattened expansion at the tip. Also the genital operculum is visibly larger on the adult male. Sexually dimorphic features cannot be seen on a subadult male. This species, the related *Typopeltis crucifer* from Taiwan, and Southeast Asian representatives (see the following species) have been hatched out in captivity. Unfortunately, *T. crucifer* has probably never made it to North America and *T. stimpsoni* imports in 2012 and 2013 likely totaled no more than a few dozen specimens (McMonigle 2013b).

Typopeltis stimpsoni is found in Japan but on just a tiny area of the mainland and a portion of the surrounding islands. The range includes the Ryukyu Archipelago, a group of small islands southwest of mainland Japan, the Satsuma Peninsula at the southern tip, and most islands just south. It has been established in the Izu Islands south of central Japan since around 1968 and is sometimes transported elsewhere with ornamental plants (Goka 2013).

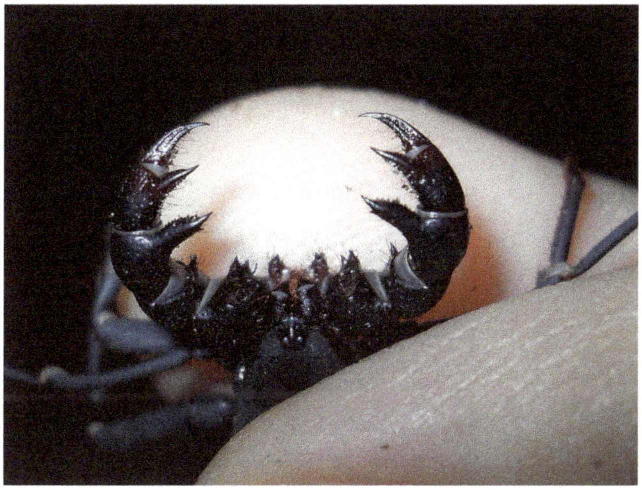

Typopeltis stimpsoni female.

Typopeltis stimpsoni, median eyes.

This species is called the Amami whipscorpion in Japan. Amami means "beautiful sky" but is also the name of a group of islands in the Ryukyu Archipelago where the species is found. There are dozens of different islands it could be named after.

Size: Adult specimens depicted here are 37 mm for the female and 43 mm for the male.

Brood Size: 30-40 (Yoshikura 1965).

Typopeltis tarnanii Pocock, 1902

Of the genera with males that possess a greatly modified tibial apophysis, only *Typopeltis* has a keel that runs from the lateral eyes towards the median eye (*Hypoctonus* and *Labochirus* do not). Although the shape of the male tibial apophysis is rather unusual, as usual, there is no sign of it on the subadult, which looks like a female. The immature male also has large dorsal spines on the trochanter which only disappear after the final molt. The male pictured here was hatched out in captivity in 2008 and did not mature until the pictured molt in 2013. Unfortunately, his siblings all succumbed to cannibalism and a brief period of dryness early on (Cain Eyre pers. comm.).

This species was available in Europe in 2008 but has probably never been on any U.S. vendor list. Most zoos that regularly import live insects like Atlas beetles and orchid mantids from Malaysia can order these in at any time from the same vendors, but they are rarely brought in except when ordered accidentally. Some of these were ordered in, and a brood was successfully hatched out, a few years ago by a U.S. insect zoo. Although this is the only *Typopeltis* species recorded for Malaysia, different imports may not represent the same species, since more than one species is found in Southeast Asia. While various specimens have mostly, or entirely, come through Malaysian exporters, it is possible they are collected in neighboring Thailand or elsewhere since collector names and locations are not divulged. Furthermore, there are other Malaysian vinegaroons from other genera, though they average smaller and the males will not have the modified apophysis. There is a slight chance this is the same as the large Malaysian specimens imported in 2000, but that species had nearly white, contrasting intersegmental membranes between the tergites. (I only had a female, so presence of a modified male tibial apophysis was unknown.)

Size: The freshly molted, captive-reared, mature male is plus or minus an inch and a half

Typopeltis tarnanii, captive-hatched subadult male. © Cain Eyre

Typopeltis tarnanii, Malaysia. Sometimes specimens look somewhat pink in photos. © Michael Seiter

This adult female imported in late 2010 was labeled "*Typopeltis crucifer* China." No males or other specimens were available.

This animal becomes obviously male, according to the pedipalps, after the final molt. *Typopeltis tarnanii* ex. Malaysia. © Cain Eyre

(~38 mm). He is a little small for this species, but his mother was larger and it is not unusual for captive-reared specimens to be smaller than wild-caught. Three wild-caught specimens acquired from a Malaysian insect house measure just under two inches (46-49 mm).

Brood Size: Approximately 30.

Typopeltis tarnanii ex. Malaysia, captive-hatched Aug. 2010, photo of immature Sept. 2013. Reared at Henry Doorly Zoo. © Trace Hardin

Typopeltis tarnanii ex. Malaysia, captive-hatched Aug. 2010, photo of immature Sept. 2013. Reared at Henry Doorly Zoo. © Trace Hardin

Typopeltis crucifer, Taiwan. © Michael Seiter

Typopeltis crucifer 2nd instar. There is a little red to the palps but there is certainly no warning coloration like that seen in *M. giganteus*. © Martin Gamache

Thelyphonellus amazonicus specimens in French Guiana. © Martin Gamache

Male, ex. Borneo. © Martin Gamache.

Female, ex. Borneo. © Martin Gamache.

Female, ex. Borneo. © Martin Gamache.

Male, ex. Borneo. © Martin Gamache.

Female, ex. Borneo. © Martin Gamache.

Male, ex. Borneo. © Martin Gamache.

GLOSSARY

Acetogenic: Capable of transforming sugars into acetic acid.

Antenniform Legs: Arachnid legs used in a wholly sensory function that do not contact the ground during walking and lack claws visible to the naked eye.

Apophysis: Fixed, usually blunt extension of the exoskeleton that serves as an attachment point for muscles.

Autotomy: Reflexive release of an appendage at a specific point of attachment in order to facilitate escape from a predator or to regenerate a damaged appendage.

Basitarsus: Basal or proximal segment of tarsus.

Book Lung: An air-filled space containing a respiratory organ filled with hemolymph (blood) that resembles a layer of flat sheets; located ventrally within the opisthosoma.

Broodsac: Extruded membranous sack containing a single clutch of eggs and attached to the female's opisthosoma; principally different from the egg sacks of insects and most arachnids because the eggs are never exposed to the atmosphere.

Captive-Bred (C.B.): Offspring produced by adults reared to maturity in captivity.

Captive-Hatched (C.H.): Offspring that result from wild-collected adult females, even if mated in captivity.

Carapace: Shield-like prosomal tergites fused into a single plate.

Cephalothorax: Prosoma; fused head and thorax common to arachnids and crustaceans.

Chelicera (*pl.* Chelicerae): Paired, pincer-like appendage in front of the mouth, unique to arachnids and other members of the subphylum Chelicerata.

Coxa (*pl.* Coxae): Basal leg segment or hip.

Cuticle: Exposed surface or outer layer of the exoskeleton.

Deimatic Display: The defensive posture of a whipscorpion that includes rearing up; threat display.

Desiccate: To dry out.

Distitarsus: The terminal or distal segment of the palp or leg tarsus.

Distal: Surface or segment furthest from the point of attachment.

Dorsal: The upper surface or top.

Entomophagus Mold: Fungus that feeds on arthropods.

Exoskeleton: The external body covering of arthropods which serves as protection and muscular support. Mainly consisting of chitin and proteins.

Exuvium (*pl.* Exuviae): The shed exoskeleton or molt. (Exuvium is common usage and makes it easy to differentiate singular and plural whereas "Exuvia" could be read as plural of exuvium or singular of exuviae.)

F1: Offspring of wild-caught adults (the offspring of two F1 would be F2 and so on).

Wild-caught would be the P or parental generation so the sequence would be P, F1, F2, F3, etc. The F stands for filial.

FEMUR (*pl.* FEMORA): Thigh; segment of an arthropod leg between the trochanter and tibia.

FLAGELLUM: Thin, threadlike appendage, normally without its own muscles, but moved by muscles near the base of attachment.

GENITAL OPERCULUM: Large second sternite containing the first pair of book lungs and gonopods which shields genitalia on the following segment. This is the first segment at the base of the opisthosoma after the pedicel.

GONOPODS: Secondary sexual organs.

HAEMOCOEL: Invertebrate body cavity through which hemolymph (blood) flows.

HYPOPUS STAGE: Suction-cup-shaped form some mite species can molt into which allows them to cling to larger invertebrates for transport to new territory.

INSTAR: Developmental stage between each molt and prior to maturity.

LATERAL: On the sides.

MEDIAN: Central or middle; the central eyes in arachnids are known as median eyes.

MEDIUM (*pl.* MEDIA): The food as well as substrate used in culturing fruit flies, bacteria, etc.

MESOSOMA: Preabdomen, widened front section of the opisthosoma where the tergites and sternites are not fused but joined by a flexible pleural membrane.

METASOMA: Postabdomen; narrow, rear section of the opisthosoma in which the tergites and sternites are fused into a rigid tube.

MOLT: The act of shedding the exoskeleton.

OCELLUS (*pl.* OCELLI): Arthropod light detection organ with a single lens.

OCULARIUM: A raised plate on the carapace which holds the median eyes and is found on many Opiliones and all Amblypygi.

OMMATOIDS: Paired dorsal, semi-translucent, ovoid areas of the exoskeleton on the last pygidial segment that may be light sensitive.

OPISTHOSOMA: Last tagma of the arachnid body; contains the book lungs and reproductive organs.

PARTHENOGENESIS: Reproduction in which unfertilized eggs develop into offspring. No vinegaroons are known to use this method, but some of the small schizomida reproduce only in this way.

PATELLA (*pl.* PATELLAE): Arachnid leg segment between the femur and tibia, "kneecap."

PEDICEL: Narrow waist (narrow first segment of the opisthosoma) connecting the prosoma and opisthosoma in certain arachnid orders including the spiders, whipspiders, and whipscorpions.

PEDIPALPS: Palps; pair of large, leg-like mouthparts located outside the chelicerae; modified into a folded trap mechanism terminating in hand-like pincers in uropygids, scorpions, pseudoscorpions, and some harvestmen.

PROSOMA: Cephalothorax; front tagma of the arachnid body; fused head and thorax from which the legs, eyes, and mouthparts originate.

PLEURAL MEMBRANE: A stretchable, lightly sclerotized membrane of the exoskeleton which joins the abdominal tergites and sternites.

POSTEMBRYO: An early, immobile developmental stage in arachnids that occurs between egg and first instar.

POSTULTIMATE MOLT: A molt that takes place after sexual maturity has been reached and after the ultimate molt.

PYGIDIUM: Postabdomen; name for the three reduced, circular segments at the end of the abdomen (similar to the five segmented scorpion metasoma or tail).

PYGIDIAL GLANDS: Paired acetogenic glands located in the thelyphonid metasoma.

SCLEROTIZE: To cause rigidity of the exoskeleton through addition of the protein sclerotin or substance other than chitin.

SETA (*pl.* SETAE): Hollow growth from the epidermal layer of the body wall; hair-like setae are often referred to as "hair" but are principally different from the hair of mammals.

SEXUALLY DIMORPHIC: Presence of noticeably different external gender characteristics.

SIGILLA: Paired indents in the sternites and tergites that serve as attachment points for muscles.

SOLUBILIZE: To make a substance dissolvable, usually in water.

SPERMATOPHORE: Structure formed by the male which includes sperm and is used in the transfer of sperm to the female.

STERNITE: Ventral abdominal plate or segment.

TAGMA (*pl.* TAGMATA): Grouping of multiple segments forming a functional group, e.g. thorax, cephalothorax, opisthosoma, and prosoma.

TARSUS (*pl.* TARSI): The foot of an arthropod. The tarsus contains a number of small segments including the claws.

TENERAL: Soft, often pale or white colored state of the exoskeleton following a molt and before it has mostly hardened or darkened (the hardening process is call sclerotization or tanning).

TELSON: Terminal segment of the scorpion's abdomen.

TERGITE: Dorsal abdominal segment.

THIGMOTACTIC: A preference to feel surfaces against the top and bottom of the body.

TIBIA (*pl.* TIBIAE): Segment of an arachnid leg between the tarsus and patella.

TRICHOBOTHRIA: Sensitive, long setae located in membranous socketed pits and commonly found in specific positions on the palps and legs, used primarily to aid in prey capture by detecting movement and vibration. The arrangement can be an important taxonomic character since they are relatively invariable.

TUBERCLE: Raised bump or short, blunt tipped spine that is part of the surface of the exoskeleton.

ULTIMATE MOLT: The final shedding of the exoskeleton from an immature and resulting in a sexually mature specimen.

Typopeltis stimpsoni hiding. Cage decor must be chosen with care so that heavy objects do not crush the whipscorpion.

BIBLIOGRAPHY

Armas, Luis F., David Ortiz, and Elier Fonseca. (2009) Depredación de huevos por dos especies de *Mastigoproctus* (Thelyphonida): no todas las presas se mueven. *Revista Ibérica de Aracnología* 17: 3-6.

Ax, Peter. (2000) *Multicellular Animals: The Phylogenetic System of the Metazoa. Volume 2.* Springer, Heidelberg, Germany.

Beccaloni, Jan. (2009) *Arachnids.* University of California Press, Berkeley, CA.

Breene, Robert G., III. (1995) Tarantulas finally catch up. *Forum of the American Tarantula Society.* 4(1): 24-26.

Breene, Robert G., III. (2000) What vinegaroons really spray. *Forum of the American Tarantula Society.* 9(4-6): 121-122.

Bruins, Eugene. (1999) *Encyclopedia of Terrarium.* Rebo Productions, Minkowsky, The Netherlands.

Carrel, James E., and Eric J. Britt. (2009) The whip scorpion, *Mastigoproctus giganteus* (Uropygi: Thelyphonidae), preys on the chemically defended Florida scrub millipede, *Floridobolus penneri* (Spirobolida: Florobolidae). *Florida Entomologist* 92(3): 500-502.

Cole, Stuart. (1998) A return to Fraser's Hill, Pahang, Malaysia. *Bulletin of the Am. Ent. Soc.* 57(417): 61-66.

Dunlop, J. A., and C. A. Horrocks. (1996) A new Upper Carboniferous whip scorpion (Arachnida: Uropygi) with a revision of the British Carboniferous Uropygi. *Zoologischer Anzeiger* 234: 293-306.

Edgecombe, Gregory D., ed. (1998) *Arthropod Fossils and Phylogeny.* Columbia University Press, New York.

Eisner, T. J., A. Meinwald, and R. Ghent. (1998) Defence mechanism of arthropods I. The composition and function of the spray of the whipscorpion, *Mastigoproctus giganteus* (Lucas), (Arachnida, Pedipalpa). *J. Insect. Physiology* 6: 272-298.

Evans, Arthur V. (2007) *Field Guide to Insects and Spiders of North America.* Sterling Publishing, New York.

Fanto, Michael. (2006) Bulldozers of the Southern US. *Invertebrates-Magazine* (Dec.): 2-6.

Franklin, Carl J. (2002) Natural history and captive maintenance of the giant vinegaroon. *Invertebrates-Magazine* (March): 7-10.

Frye, Frederic L. (1992) *Captive Invertebrates: A Guide to Their Biology and Husbandry.* Krieger Publishing, Malabar, FL.

Geethbali and S. D. Moro. (1988) The disposition and external morphology of trichobothria in two arachnids. *Acta. Arachnol.* 36: 11-23.

Goka, K. (2013) *Typopeltis stimpsonii.* Invasive Species of Japan. (website) National

Institute for Environmental Studies, Japan. http://www.nies.go.jp/biodiversity/invasive/DB/detail/70500e.html

Harvey, Mark S. (2003) *Catalogue of the Smaller Arachnid Orders of the World: Amblypygi, Uropygi, Schizomida, Palpigradi, Ricinulei and Solifugae.* CSIRO Publishing, Collingwood, Victoria, Australia.

Harvey, Mark. (2011) Order Pseudoscorpiones de Geer, 1778. *In*: Zhang, Z.-Q. (Ed.) Animal biodiversity: An outline of higher-level classification and survey of taxonomic richness. *Zootaxa* 3148: 120.

Haupt, J. (2009) Proposal for the synonymy of some South-East Asian whip scorpion genera (Arachnida, Uropygi, Thelyphonida). *Rev. Iber. Aracnol.* 17: 13-20.

Haupt, J., and F. Mueller. (2004) New products of defense secretion in South East Asian whip scorpions (Arachnida: Uropygi: Thelyphonida). *Mem. Soc. Ent. Italy* 78(2): 305-319.

Haupt, J., et al. (1988) Chinese whip scorpion using 2-ketones in defense secretion (Arachnida: Uropygi). *Journal of Comparative Physiology B* 157(6): 883-885.

Hernandez, Spencer. (2001) An introduction to whipscorpions. *Arachnida* 1: 8-11.

Huff, Jeremy C., and Lorenzo Prendini. (2009) On the African whip scorpion, *Etienneus africanus* (Hentschel, 1899) (Thelyphonida, Thelyphonidae), with a redescription based on new material from Guinea-Bissau and Senegal. *American Museum Novitates* (3658): 1-16.

Itokawa, Hideki, Kano Rokuro, and Nakajima Terumi. (1985) Chemical investigation of the spray of the Asian whipscorpion *Typopeltis stimpsoni* (Wood, 1862). *Medical Entomology and Zoology* 36(1): 65-66.

Krehenwinkel, H., et al. (2009) On *Thelyphonoides panayensis* gen. et sp. n. (Arachnida: Uropygi: Thelyphonidae), a new genus and a new species of whip scorpions from Panay Island (Philippines). *Arthropoda Selecta* 18(3-4): 139-143.

Levi, Herbert W., and R. Lorna. (1968) *Spiders and Their Kin.* Western Publ., Racine, WI.

McGavin, George C. (2000) *Insects, Spiders, and Other Terrestrial Arthropods.* Dorling Kindersley, London.

McMonigle, Orin. (2006) *Giant Tarantulas: The Enthusiast's Handbook.* Elytra and Antenna, Brunswick, Ohio.

McMonigle, Orin. (2008) *Whipscorpions and Whipspiders: Culturing Gentle Monsters.* Elytra and Antenna, Brunswick, Ohio.

McMonigle, Orin. (2009) *Scorpions, Windscorpions, Pseudoscorpions: Culturing Ancient Arachnids.* Elytra and Antenna, Brunswick, Ohio.

McMonigle, Orin. (2013) *Breeding the World's Largest Arachnid: Amblypygid Biology and Natural History.* Coachwhip Publications, Greenville, Ohio.

McMonigle, Orin. (2013b) The Ninjaroon, Japan's spectacular whipscorpion *Typopeltis stimpsoni* (Wood, 1862). *Invertebrates-Magazine* 12: 4.

Milne, Lorus, and Margery Milne. (1980) *National Audubon Society Field Guide to North American Insects and Spiders.* Knopf Publishers, New York.

O'Toole, Christopher, ed. (2002) *Firefly Encyclopedia of Insects and Spiders.* Firefly Books Ltd., Buffalo, NY.

Prendini, Lorenzo. (2011a) Order Schizomida Petrunkevitch, 1945. *In*: Zhang, Z.-Q. (Ed.) Animal biodiversity: An outline of higher-level classification and survey of taxonomic richness. *Zootaxa* 3148: 156.

Prendini, Lorenzo. (2011b) Order Solifugae Sundevall, 1833. *In*: Zhang, Z.-Q. (Ed.) Animal biodiversity: An outline of higher-level classification and survey of taxonomic richness. *Zootaxa* 3148: 118.

Prendini, Lorenzo. (2011c) Order Amblypygi Thorell, 1883. *In*: Zhang, Z.-Q. (Ed.) Animal biodiversity: An outline of higher-level classification and survey of taxonomic richness. *Zootaxa* 3148: 154.

Prendini, Lorenzo. (2011d) Order Thelyphonida Latreille, 1804. *In*: Zhang, Z.-Q. (Ed.) Animal biodiversity: An outline of higher-level classification and survey of taxonomic richness. *Zootaxa* 3148: 155.

Punzo, Fred. (1998) *The Biology of Camel-Spiders (Arachnida, Solifugae)*. Kluwer Academic Publishers, Norwell, MA.

Punzo, Fred. (2000) Diel activity patterns and diet of the giant whipscorpion *Mastigoproctus giganteus* (Lucas) (Arachnida, Uropygi) in Big Bend National Park (Chihuahuan Desert). *Bull. of the British Arachnol. Soc.* 11(9): 385-387.

Punzo, Fred. (2005) Habituation, avoidance learning, and spatial learning in the giant whipscorpion, *Mastigoproctus giganteus* (Lucas) (Arachnida, Uropygi). *Bull. of the British Arachnol. Soc.* 13(4): 138-144.

Punzo, Fred. (2006) Types of shelter sites used by the giant whipscorpion *Mastigoproctus giganteus* (Arachnida, Uropygi) in a habitat characterized by hard adobe soils. *Journal of Arachnology* 34: 266-268.

Punzo, Fred, and C. Reeves. (2001) Geographical variation in male courtship behaviour of the giant whipscorpion *Mastigoproctus giganteus* (Lucas) (Arachnida, Uropygi). *Bull. of the British Arachnol. Soc.* 12(2): 93-96.

Roth, Luis M. (2002) Cockroach reproduction and phylogeny. *Invertebrates Magazine* 2(1): 2-9.

Rowland, Mark J., and John A. Cooke. (1973) Systematics of the arachnid order Uropygida (=Thelyphonida). *Journal of Arach-nology* 1: 55-71.

Santer, Roger D., and Eileen A. Hebets. (2007) Agonistic signals received by an arthropod filiform hair allude to the prevalence of near-field sound communication. *Proceedings of the Royal Society* 275: 363-368.

Sato, Ikio. (1941) Some ecological notes on the whip-scorpion, *Typopeltis stimpsoni*. *Acta Arachnol.* 6: 72-87.

Sejna, Vladimir. (2005) Schizomida. *Sklipkan* 1: 32.

Schmidt, Justin O. (1998) *Courtship, Female Choice and Male Investment in the Vinegaroon, Mastigoproctus giganteus.* ISA presentation XIV.

Schmidt, Justin O. (2000) *Ecology and Life History Strategies of the Vinegaroon.* Creative Cooperation in Resource Management III, presentation.

Schmidt, Justin O. (2003) Vinegaroon (*Mastigoproctus giganteus*) life history and rearing. *Proc.* 2003 *Invertebrates in Captivity Conf.*, pp. 73-80. Sonoran Arthropod Studies Institute: Tucson, AZ.

Schmidt, Justin O. (2009) Vinegaroons. in *Encyclopedia of Insects*. 2nd ed. by Resh, V. H., and R. T. Carde. Academic Press, Burlington, MA.

Schmidt, Justin O., and Li S. Schmidt. (1998) *Ecology and Population Biology of the Vinegaroon, Mastigoproctus giganteus.* ISA presentation XIV.

Schmidt, J., et al. (1998) *Chemical Composition of the Allomones of Adult and Immature Vinegaroons Mastigoproctus giganteus.* International Society of Chemical Ecology Presentation, 15th Ann. Mtg., Ithaca, NY.

Schmidt, J., et al. (2000) Chemistry, ontogeny, and role of pygidial gland secretions of the vinegaroon *Mastigoproctus giganteus*

(Arachnida: Uropygi). *Journal of Insect Physiology* 46(4): 443-450.

Shultz, Jeffrey W. (1992) Step-coupled fluctuations in prosomal pressure may constrain stepping rates in whipscorpions (Uropygi). *Journal of Arachnology* 20: 148-150.

Sim, J. H., et al. (2007) *Clostridium aceticum*—a potential organism in catalyzing carbon monoxide to acetic acid: Application of response surface methodology. *Enzyme and Microbial Technology* 40(5): 1234-1243.

Stoops, Erik D., and Jeffrey Martin. (1995) *Scorpions and Venomous Insects of the Southwest*. Golden West Pub., Phoenix, AZ.

Blanford, C. T., ed. (1900) *The Fauna of British India, Including Ceylon and Burma*. Taylor and Francis, Calcutta, Bombay.

Viquez, Carlos. (2003) Whip scorpions of Central America. *Sklipkan Journal for Invert. Keepers and Breeders* 8(3): 86.

Viquez, C., and L. F. Armas. (2005) Dos nuevos generos do vinagrillos do Centroamerica y las Antillas (Arachnida: Thelyphonida). *Boletin de la SEA* 37: 95-98.

Viquez, C., and L. F. Armas. (2006) Un nuevo genero y dos nueva especies de vinagrillos Centroamericanos (Arachnida: Thelyphonida). *Boletin do la SEA* 38: 37-41.

Werner, F., and C. Olson. (1994) *Insects of the Southwest*. Fisher Books, Tuscon, AZ.

Wheeler, Ward C., and Cheryl Y. Hayashi. (1998) The phylogeny of the extant chelicerate orders. *Cladistics* 14: 178-192.

Weygoldt, Peter. (2000) *Whip Spiders (Chelicerata: Amblypygi): Their Biology, Morphology and Systematics*. Apollo Books, Denmark.

Weygoldt, Peter, and Siegfried Huber. (2013) Sperm transfer and maternal care in *Thelyphonus* c.f. *caudatus* from Lombok, Indonesia (Arachnida, Uropygi, Thelyphonida). *Zoologischer Anzeiger*. 252(3): 348-349.

Yogi, S., and J. Haupt. (1977) Analyse des wehersekretes bei dem geibelscorpion *Typopeltis crucifer* Pocock, 1894. *Acta Arachnol*. 27: 53-56.

Yoshikura, M. (1965) Postembryonic development of a whip-scorpion, *Typopeltis stimpsonii* (Wood). *Kumamoto J. Science, Ser. B* 7(2): 21-51.

Typopeltis stimpsoni. The first tergite is very small and often overlapped by the prosoma.

COACHWHIP PUBLICATIONS
COACHWHIPBOOKS.COM

INVERTEBRATES FOR EXHIBITION
ORIN MCMONIGLE
ISBN 978-1-61646-147-8

MILLIPEDS IN CAPTIVITY
ORIN MCMONIGLE
ISBN 978-1-61646-143-0

THE ULTIMATE GUIDE TO BREEDING BEETLES
ORIN MCMONIGLE
ISBN 978-1-61646-132-4

KEEPING THE PRAYING MANTIS
ORIN MCMONIGLE
ISBN 978-1-61646-166-9

PILLBUGS AND OTHER ISOPODS
ORIN MCMONIGLE
ISBN 978-1-61646-207-9

BREEDING THE WORLD'S LARGEST LIVING ARACHNID
ORIN MCMONIGLE
ISBN 978-1-61646-183-6

INVERTEBRATES
FOR EXHIBITION
ORIN MCMONIGLE
INSECTS, ARACHNIDS, AND
OTHER INVERTEBRATES SUITABLE
FOR DISPLAY IN CLASSROOMS,
MUSEUMS, AND INSECT ZOOS

MILLIPEDS IN CAPTIVITY
DIPLOPODAN HUSBANDRY AND
REPRODUCTIVE BIOLOGY
ORIN A. MCMONIGLE

THE ULTIMATE GUIDE
TO BREEDING BEETLES
COLEOPTERA LABORATORY CULTURE METHODS
ORIN MCMONIGLE

KEEPING THE
PRAYING MANTIS
Orin A. McMonigle

PILLBUGS AND
OTHER ISOPODS
CULTIVATING VIVARIUM CLEAN-UP CREWS AND
FEEDERS FOR DART FROGS, ARACHNIDS, AND INSECTS
Orin McMonigle

BREEDING THE WORLD'S
LARGEST LIVING ARACHNID
AMBLYPYGID BIOLOGY, NATURAL HISTORY, AND CAPTIVE HUSBANDRY
ORIN MCMONIGLE